Food Waste to Animal Feed

Food
Waste

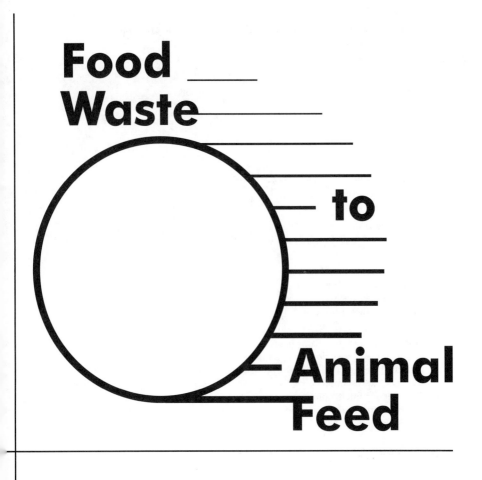

to

Animal
Feed

Edited by   Michael L. Westendorf

   Iowa State University Press / Ames

Michael L. Westendorf, PhD, is an Extension Animal Scientist, Cook College, Rutgers University.

Iowa State University Press
2121 South State Avenue, Ames, Iowa 50014

Orders: 1-800-862-6657
Office: 1-515-292-0140
Fax: 1-515-292-3348
Web site: www.isupress.com

♾Printed on acid-free paper in the United States of America

First edition, 2000

Library of Congress Cataloging-in-Publication Data

Food waste to animal feed / edited by Michael L. Westendorf.— 1st ed.
    p. cm.

Includes bibliographical references (p.).
    ISBN 0–8138–2540–7 (alk. paper)
    1. Food waste as feed—Congresses. 2. Animal feeding—Congresses.
    I. Westendorf, Michael L.

SF99.F67 F66 2000
636.08'556—dc21

00–059666

The last digit is the print number: 9 8 7 6 5 4 3 2 1

F
7
67
F66
2000

Contents

Appendixes

Contributors

(Chapter numbers appear in parentheses.)

Bonnie A. Altizio, MS (5)
Extension Program Associate
Rutgers University
New Brunswick, NJ

Joel H. Brendemuhl, PhD (8)
University of Florida
Gainesville, FL

Barbara Corso, DVM, MS (7)
USDA, APHIS, VS, CEAH
Fort Collins, CO

Perry J. Durham, DVM (14)
Manager of Quality, Feed Division
Farmland Industries
Kansas City, MO

Barbara L. Ferko-Cotten, DVM (10)
North Carolina State University
Raleigh, NC

Don A. Franco, DVM, MPH (13)
Director of Scientific Services
National Renderers Association, Alexandria, VA
    and Animal Protein Producers Industry
Huntsville, MO

Harold W. Harpster, PhD (9)
Pennsylvania State University
University Park, PA

Roger D. Hoestenbach (4)
Head, Feed and Fertilizer Control Service
Texas Agricultural Experiment Station
College Station, TX

Dwain D. Johnson, PhD (8)
University of Florida
Gainesville, FL

Daniel G. McChesney, Ph.D. (3, 12)
Deputy Director
Office of Surveillance and Compliance
FDA, Center for Veterinary Medicine
Rockville, MD

Robert O. Myer, PhD (8)
University of Florida
Marianna, FL

Gary Pearl, DVM (13)
President
Fats and Proteins Research Foundation
Bloomington, IL

Matthew H. Poore, PhD (10)
North Carolina State University
Raleigh, NC

Glenn M. Rogers, DVM, MS, Dipl. ABVP (10)
North Carolina State University
Raleigh, NC

Jonathan R. Schultheis, PhD (10)
North Carolina State University
Raleigh, NC

Felix J. Spinelli, PhD (7)
Agricultural Economist
Food Safety and Inspection Service, USDA
Washington, D.C.

Arnold C. Taft, DVM (5)
Senior Staff Veterinarian, USDA
Animal and Plant Health Inspection Service
Riverdale, MD

Paul Walker, PhD (2, 11)
Illinois State University
Normal, IL

Michael L. Westendorf, PhD (1, 6)
Extension Animal Scientist
Rutgers University
New Brunswick, NJ

Ernest W. Zirkle, DVM (5)
Director, Division of Animal Health
New Jersey Department of Agriculture
Trenton, NJ

Preface

The initial motivation for this book originated several years ago when it was suggested by Dr. Arnold Taft of the USDA Animal and Plant Health Inspection Service that a symposium be held on the use of food waste as animal feed. Five symposia have since been held to discuss this idea. These symposia were sponsored by the New Jersey Department of Agriculture, Rutgers Cooperative Extension, USDA-APHIS, and the National Pork Producers Council. The organization that resulted from these meetings is now called the Food Recovery and Recycling Association of North America.

These meetings have focused upon research and product development, equipment and new technology, regulation and terminology, and the food-waste dilemma. The magnitude of the food-waste disposal problem cannot be understated. According to a Franklin and Associates (1998) survey conducted for the Environmental Protection Agency, food waste contributes 10%, or 21.9 million tons, of municipal solid waste (MSW). Only about 2.4% of this total is recycled. Food waste, because of the low disposal rate and other problems associated with disposal, may represent the single greatest disposal problem for MSW.

Food waste's high nutrient content makes it a potential animal feed. Most analyses reveal food waste to have high protein and fat content, both in excess of 20%. Any animal feeding problems relate primarily to animal health concerns, moisture content, and nutrient variability. The bulk of research completed with food waste has used wet waste for animal feed; however, recent projects have used various processed (extruded, dehydrated, pelleted, ensiled, etc.) products in animal feeding experiments. The ability to further process and dewater food waste would allow preservation, storage, and easier use commercially.

Terminology varies throughout the book. Some authors use the term food waste to refer to food plate waste only, while others may use it to refer to either plate or nonplate food waste (plate waste being unique because it requires further cooking or processing to meet mandated federal sanitation requirements). The terms food residual or recycled food commodity also are used. My own definition is that food waste refers to products such as plate waste, but also includes supermarket waste (meat trimmings, produce, deli waste), food processing waste (especially vegetable waste), fish and cannery waste, and various types of bakery waste. These products are not normally considered to be by-product animal feeds. There are numerous by- or coproducts of other industries currently fed to animals, examples being brewers and distillers grains, beet pulp, citrus pulp, soy hulls, and cottonseed, to name a few. These have been fed for many years, are consistent in nutrient content, and are often available regionally, if not nationally. Food waste is not consistent in quality, is usually high in moisture content, and only available locally. It is usually free for the taking and farmers are often paid to remove it.

This food waste is not only difficult to dispose of, but it is also a challenge to process as animal feed. This book focuses on the challenges of utilizing both wet and/or processed food waste. The regulatory environment relating to food waste, the perspective of the end-users, and practical use as animal feed is also discussed. The first five chapters give an overview and focus on the regulatory aspects of food waste use. There are three chapters on the use of food waste as a swine feed and three more on its use as a ruminant feed source. The final chapters focus on end use and some concerns of people in the industry.

I would like to thank Drs. Arnold Taft of the USDA-APHIS and Ernie Zirkle of the New Jersey Department of Agriculture, Division of Animal Health, for their work on this subject. Drs. Beth Lautner and Paul Sundberg of the National Pork Producers Council also have been helpful. Thanks to Drs. Pat Schoknecht, Harold Hafs and Donald Derr of Rutgers University, and to Drs. George Mitchell and Virgil Hays of the University of Kentucky for their assistance with reviewing.

One of the goals of this publication, other than to give a clear explanation of the subject, was to stimulate a need for research. This is an issue with great possibilities. Those who recycle food waste often gain credits within their own states while helping food waste generators to dispose of their food waste burden. The USDA also has discussed new initiatives for those who recycle food waste. Finally, food waste can make a good, economical food source if processed correctly.

Food Waste to Animal Feed

1

Food Waste as Animal Feed: An Introduction

by Michael L. Westendorf

Feeding food waste to animals long has been an important component of livestock production and provides a competitive alternative to more traditional feedgrain or protein sources. Some feedstuffs that are by-products or coproducts of other industries (corn gluten meals and feeds, wheat middlings, brewers or distillers grain) are no longer considered food waste by-products but rather normal commodities to be included in livestock diets. For the purpose of this book, food waste refers to any by-product or waste product from the production, processing, distribution, and consumption of food. These products can include cull beans and potatoes, vegetable or fruit waste from a food processor, dairy processing waste, bakery waste, and garbage. Food waste normally may be disposed of in landfills, in incinerators, or by land application. It is often high in moisture content, while nutrient content is usually adequate, although highly variable. As an example, garbage averages in excess of 20% protein and 20% fat but may be only 25% dry matter and have a coefficient of variation (CV) of 40% for protein and 75% for fat. Thus, although these feeds have nutritive value, they are difficult to incorporate into a commercial feeding program.

One of the earlier recorded uses of garbage as an animal feed was described by Minkler (1914). Figure 1.1 and figure 1.2 picture garbage-fed hogs earlier this century. Throughout the state of New Jersey, farmers were feeding chiefly hotel and resort waste to pigs. In the vicinity of Secaucus, in Hudson County, at any time there were more than 25,000 pigs fed entirely hotel food waste, obtained from New York City, Newark, and Jersey City. In other parts of the state, in addition to food waste, pigs were fed skim milk and buttermilk, perishable fruits and vegetables, and feed milling or grain wastes. Table 1.1 shows results of an early garbage-feeding experiment conducted at the New Jersey Agricultural Experiment Station (NJAES 1919). This experiment compared cooked

3

Figure 1.1 Garbage-fed hogs. Taken from Minkler (1914).

Figure 1.2 Group of hogs consuming cooked garbage. Taken from Kornegay et al. (1970).

Table 1.1. Performance of pigs fed garbage

Lot	Ration	Initial Wt. (lbs)	Final Wt. (lbs)	Total Gain per Pig (lbs)	Garbage Consumed per Pig (lbs)	Grain Consumed	Garbage per 100 lb Gain	Grain per 100 lb Gain
1	Cooked garbage	30.4	195.2	165.1	4022.3	—	2436.9	—
2	Raw garbage	30.4	219.8	189.4	4249.2	—	2243.6	—
3	Raw garbage (finished on grain)	25.4	245.3	219.9	2936.0	291.7	1335.2	132.6
4	Raw garbage plus 1% shelled corn	33.9	291.9	258.0	4237.8	281.9	1642.6	109.3
5	Grain ration	33.1	245.0	211.9	—	861.8	—	406.7
6	Raw garbage plus forage	30.8	200.9	170.2	4163.3	—	2446.8	—

Note: Treatments consisted of Lot 1, cooked garbage; Lot 2, raw garbage; Lot 3, raw garbage (finished last 40 days on grain); Lot 4, raw garbage plus 1 % shelled corn; Lot 5, grain ration (shelled corn, wheat middlings, and tankage); Lot 6, raw garbage plus green succulent forage. Ten pigs per treatment.
Source: NJAES 1919.

garbage, raw garbage, a grain ration, and several combinations of raw garbage fed and supplemented. Results indicated that supplementing food waste with grain improved performance to a level similar to the performance of pigs fed grain. Figure 1.3 shows some of the meat resulting from this experiment. There is still concern about the quality of meat from animals fed food waste (Westendorf et al. 1998). Numerous other experiments over the past 75 years have shown that proper supplementation improves the performance of food waste-fed pigs (even when food waste is of very poor quality) to levels similar to grain-fed animals (Kornegay et al. 1970; Westendorf et al. 1998; Altizio et al. 1998).

Much of this research was completed prior to enactment of the Swine Health Protection Act (U.S. Congress 1980) that established rules for food-waste feeding. Nevertheless, many of these early studies recognized that the cooking of food waste (garbage) was important (Henry and Morrison 1920). Food waste was often cooked in rendering vats to extract the grease, which was sold separately (Minkler 1914; NJAES 1919). The profit from selling the grease paid for collecting and treating the garbage, leaving income from the pigs as profit (Minkler 1914; Henry and Morrison 1920). Some food waste or by-products other than garbage that were fed to swine earlier this century, are listed in Table 1.2. It is interesting to note that there was apparently (Minkler 1914) no

Figure 1.3 Hams and bacon from hogs in a garbage-feeding experiment. Taken from NJAES (1919).

Table 1.2. Food wastes fed to pigs

Skim milk	Cull bean
Buttermilk	Dried distillers grains
Whole milk	Distillery slop
Whey	Meat meal
Tankage	Blood meal

Source: Henry and Morrison 1920.

plan for manure waste management for the 25,000 pigs housed outside Secaucus, and all waste was "flushed into open gutters . . . that led off into the tidal meadows."

By the 1960s, there were no food waste or garbage swine feeders remaining in the Secaucus area, and the food-waste or garbage-feeding industry became consolidated in southern New Jersey (Koch 1964), as it is today. Food waste fed today is primarily institutional (hospital, nursing home, prison) and restaurant waste (Westendorf et al. 1996). It may often be supplemented with bakery, fish, or vegetable wastes. In the period from the 1960s until 1994, the number of state-licensed swine food-waste feeders declined from 250 to 36, and the number of pigs finished declined from 130,000 head finished annually to less that 50,000 (Westendorf et al. 1996).

Different forms of processed garbage also have been used in the diets of ruminants. Food waste processed to remove glass, metal, and plastic and then dehydrated has been fed to both beef cattle and sheep. Cattle performed well when processed food waste was incorporated into the diet (McClure et al. 1970), with acceptable digestibility and palatability. McClure et al. (1970) also found that the intake of processed food waste in both cattle and sheep was similar to more traditional diets and concluded that dehydrated, processed food waste might be comparable to or competitive with other feedstuffs such as hay.

Solid Waste

Food waste comprises approximately 10% or about 22 million tons of the total municipal solid waste stream (Franklin and Associates 1998) and is the least likely to be recycled (Table 1.3). Only 2.4% of the food waste produced is recovered or recycled. About 20% of all food produced for human consumption is wasted in production, processing, packaging, distribution, and consumer waste (Tolan 1983).

According to Franklin and Associates (1998) food wastes considered part of municipal solid waste "consist of uneaten food and food preparation wastes from residences, commercial establishments (restaurants and fast-food establishments), institutional sources such as school cafeterias, and industrial sources such as factory lunchrooms." Food waste

Table 1.3. Materials generated in municipal solid waste weight by weight, 1996

	Million Tons	% of Total	% Recycled
Paper and paperboard	79.9	38.1	40.8
Yard trimmings	28.0	13.4	38.6
Plastics	19.8	9.4	5.3
Metals	16.1	16.1	39.6
Wood	10.8	5.2	4.5
Food waste	21.9	10.4	2.4
Glass	12.4	5.9	25.7
Other	20.8	9.9	11.5
Total	209.7		

Source: Franklin and Associates 1998.

generated during preparation and packaging of food products is considered industrial waste and not part of the estimates in table 1.3. The 21.9 million-tons represent a substantial increase over previous estimates (Franklin and Associates 1998), possibly due to curbside recycling estimates that are higher than previously.

Food Processing

There is a great amount of food waste produced other than municipal solid waste as described by CAST (1995) and the NRC (1983). The best data available (CAST 1995) for wastes produced from fruit and vegetable processing estimate that all post-harvest losses during handling, processing, and packaging total nearly 7.5 million tons.

Waste solids separation accumulates particulate solids that may be useful in animal feed rations. These products can be converted into dry or pelleted food waste and may be sold in bulk as animal feed. According to CAST (1995), in 1971 there were 5,000 dairy-processing plants producing 53 billion gallons of wastewater annually, 31 billion gallons discharged directly into watercourses. CAST (1995) has described new techniques in the milk waste processing industry. Ultrafiltration now can separate biological material. Whey is the largest (Clark 1979) waste from the industry. In 1976, an estimated 15.6 million tons of liquid whey were produced or over 1 million tons of whey solids. Whey can be fed to both swine and ruminants, can be incorporated into milk replacers, and will continue to be used as an animal feed.

Wastes from meat and poultry processing have been and will continue to be used as animal feeds. CAST (1995) described the wastes from beef or poultry as primarily blood, feathers, bone, and offal. Blood, feathers, and bone usually are processed into a meal product for animal feed. Meat scraps unsuitable for human consumption are sold or given to rendering facilities for processing into animal and pet foods. The raw materials used are composed of packing house offal and bone, dead stock (whole animals), butcher shop fat and bone, blood, restaurant grease, and possibly offal and feathers. A 1,000-pound steer will produce approximately 600 pounds of edible product (includes lean meat, fat, and bone) and the remaining 400 pounds will be processed at a rendering plant.

In 1994, renderers produced 8.4 billion pounds of tallow and grease, one-third of which was fed to animals (Prokop 1996). Six billion pounds of meat meal, feather meal, and other protein products also were produced with 90% of these protein meals fed to cattle, hogs, and poultry. The remaining 36 billion pounds of available by-product waste was mostly inedibles and water. When not rendered within 24 hours, these materials become a liability (Prokop 1996). Rendering these materials is a benefit both environmentally and economically, as they are marketed in competition with other animal fats, vegetable oils and proteins.

Economics is another challenge when rendering meat by-products or processing food wastes (garbage) that will be discussed later. It is costly to dehydrate and process high-moisture products. These products will compete in the commodities markets with other sources of complete feeds, fats and proteins, and must be cost competitive.

CAST (1995) reported that the disposal options for the seafood industry are as meal (fish or crab meal), pet food, rendering, composting, landfilling, ocean dumping, and bait. Currently, the seafood industry (Pigott 1981) thinks of seafood waste as a secondary raw material having a variety of uses, helping to alleviate the seafood waste problem.

In 1989, there were 311,199 tons of fish meal produced for animal feed (USDA-ERS 1992). Most of this was produced from menhaden fish as a by-product of oil extraction. Processing fish products (CAST 1995) as fertilizer, composting, and the use of fish protein hydrolysates as animal feeds are some of the new potential uses for fish by-products.

Yield of edible products can illustrate one of the problems with processing and the challenges and/or opportunities to use seafood wastes as animal feed. Whole cod yields represent about 47% of total weight with 53% waste while salmon and perch yield 64 and 33%, respectively. Shellfish yields are generally less. Blue crab yields are between 10 and 15% with 85 to 90% waste, while shrimp, scallop, and oysters average 30 to 40, 10 to 18, and 11 to 17% yield respectively (Waterman 1975). For a population that consumed nearly 4 billion pounds of fish in 1996 (USDA 1997), this represents a tremendous amount of potential seafood, fish, and shellfish waste. Fontenot et al. (1982) compared fish and crab nutrient content. Both are high in protein; however, crab is much higher in ash and calcium than fish waste and also may serve as a mineral source. Oyster shells are used as a mineral supplement and shrimp also has elevated mineral levels (Ensminger et al. 1990). Wohlt et al. (1994) found that feeding sea-clam viscera as a protein supplement to pigs had no effect on carcass characteristics, although there were palatability differences. New means of processing fish and shellfish wastes may yield new animal feeds.

Animal Feed Composition

Table 1.4 lists a number of food processing by-products that are used as animal feeds. It is interesting that most are no longer considered (USDA-ERS 1992) as waste by-products but rather as commodities to be incorporated in complete feeds. A thorough survey of the nutrient content of alternative feeds was completed by Bath et al. (1998) and is available from Feedstuffs®.[1] Studies by Kornegay and colleagues (Barth et al. 1966; Kornegay et al. 1968) analyzed four sources of food waste: hotel and restaurant, institutional, military, and municipal food waste (table 1.5). Feeding trials indicated that grain supplementation improved the performance of food waste-fed pigs (Kornegay et al. 1970). Since municipal waste has the poorest quality, supplementation benefits it the most. Recent studies (Westendorf et al. 1999) with food waste from several sources (see chapter 6) indicated that today's food waste has a nutrient content similar to that fed 30 years ago—high in protein and fat. Although it has ash levels lower than 30 years ago, most minerals are near adequate for the majority of livestock species.

Table 1.4. Food processing by-products - type and amount produced

Type	Amount (in tons)
Animal Products	
Meat and bone meal	2.0 million
Edible tallow	2.8 million
Restaurant grease	1.1 million
Feather meal	200,000
Fish meal	311,199
Dry whey	98,477
Dry milk	9,372
Grain Milling Byproducts	
Corn gluten feed and meal	6.4 million
Distillers grains	1.4 million
Brewers dried grains	117,300

Source: USDA-ERS 1992.

Table 1.5. Nutrient content of waste fed to experimental pigs

| Nutrient [a] | Source of Waste | | | |
	Hotel/ Restaurant	Institutional	Military	Municipal
Dry matter (%)	16.00[c]	17.50[c]	25.60[b]	16.60[c]
Crude protein (%)	15.30[c]	14.60[c]	15.90[b,c]	17.50[b]
Ether extract (%)	24.90[c]	14.70[d]	32.00[b]	21.40[c]
Crude fiber (%)	3.30[c]	2.80[c]	2.80[c]	8.40[b]
Ash (%)	5.70[c]	5.20[c]	5.50[c]	8.60[b]
Calcium (%)	0.47[c]	0.37[c]	0.47[c]	1.69[b]
Phosphorous (%)	0.35	0.29	0.36	0.39
Copper (ppm)	34.00	27.00	45.00	50.00
Iron (ppm)	510.00	429.00	435.00	346.00
Magnesium (ppm)	966.00	548.00	694.00	766.00
Manganese (ppm)	24.00	23.00	36.00	30.00
Carotene (ppm)	4.40	4.20	3.90	4.90
Thiamine (ppm)	2.20	2.70	2.30	3.90
Riboflavin (ppm)	3.40	2.90	2.70	3.80
Niacin (ppm)	27.90	25.20	29.90	23.00
Pantothenic acid (ppm)	2.70[c]	4.30[c]	3.90[c]	6.20[b]
Total digestible nutrients (as fed) (%)	15.50	19.40	33.60	15.90

Source: Kornegay et al. 1968, 1970; Barth et al. 1966.
[a]All nutrients reported on a dry matter basis.
[b,c,d]Means in a row with different superscripts differ ($p<0.05$).

High moisture content is the main difficulty when incorporating food waste into practical feeding programs. This requires that food waste be fed immediately to limit spoilage or further processing to decrease moisture content. Additionally, food waste has high coefficients of variation (CVs). While the energy and protein levels indicate that food waste has excellent nutritive value, the high CVs make it difficult to incorporate food waste in current feeding programs because this variability makes ration balancing difficult.

Animal Health

The issue of animal health is always a concern when discussing the use of food waste as animal feed. When food waste is fed wet it has only a very brief shelf life, and there is always the risk of foreign animal diseases (USDA-APHIS 1990). Outbreaks of disease have been associated with the feeding of uncooked garbage and ultimately led to the requirement of cooking food waste. Hog cholera (Minkler 1914) was a concern for producers earlier in this century, although not directly associated with food-waste feeding at the time. Hog cholera and other foreign animal diseases were the catalyst leading to the prohibitions and restrictions on the use of food waste or garbage as an animal feed, as expressed in the federal Swine Health Protection Act, which passed in 1980. This Act dictated that all food waste be cooked, specifically "all waste material derived in whole or in part from the meat of any animal (including fish and poultry) or other animal material, and other refuse of any character whatsoever that has been associated with any such material, resulting from the handling, preparation, cooking, or consumption of food, except that such term shall not include waste from ordinary household operations that is fed directly to swine on the same premises where such household is located." It is the presence of meat that necessitates cooking, whether that meat originates from table scraps or is a by-product of food preparation. All table or plate scraps require cooking before feeding to swine (except for those produced and fed upon household premises). The Act, as amended above, does not require the cooking of non-meat food processing or by-product items.

The foreign animal diseases of concern include hog cholera, foot and mouth disease, African swine fever, and swine vesicular disease. There also is concern about domestic pathogens of public health significance such as *Salmonella, Campylobacter, Trichinella,* and *Toxoplasma.* The primary concern with foreign animal diseases is the importation of both legal and contraband (illegally imported) materials that may ultimately be discarded and be fed as food waste. A USDA-APHIS (1995) survey dealt with this subject and will be described in greater detail in Chapter 6. Nevertheless, the requirement for cooking still includes all food and plate waste fed to swine.

There is no specific requirement for cooking food waste for feeding ruminants. However, recent rules enacted because of the risk of Bovine Spongiform Encephalopathy require that food waste, which includes ruminant product, must be processed prior to feeding (see Chapter 3).

Conclusion

Until the last several years, most food waste research has been conducted with wet food waste (Kornegay et al. 1970; Westendorf et al. 1996; Westendorf et al. 1998). New processing technologies to produce a drier product may make it easier to include food wastes in commercial diets, to reduce product variability, and, ultimately, to increase the level of food-waste recycling. Myer et al. (1994, 1999), Rivas et al. (1994), and Altizio et al. (1998) have all conducted research with processed food waste fed to pigs. In addition, Walker and Kelly (1997) has done work with processed food waste as a feedstuff for both ruminants and mono-gastric animals. All of these involve some form of further processing (extruding, pelleting, or dehydrating) or some other treatment such as ensiling.

This book focuses on the use of food wastes as animal feed. Issues of safety, regulation, management, processing, and quality control will be discussed. It is hoped that this discussion will stimulate further interest in the topic. It is unclear just how much food waste is available. The esti-mates used in this chapter are from several different sources (EPA, USDA, CAST, NRC, and several private estimates) and may often appear to be inconsistent comparisons. Estimates are that 20% of food from production to consumption is lost as waste. The 21.9 million ton figure given earlier only represents municipal waste and not that from food processing or rendering waste. These numbers indicate that improve-ments in the recycling of food waste are possible. The use of food waste as animal feed may optimize energy savings by comparison with the other options (landfills, incinerators, biosolids, soil amendments, etc.) currently available.

References

Altizio, B. A., P. A. Schoknecht, and M. L. Westendorf. 1998. Growing swine pre-fer a corn/soybean diet over dry, processed food waste. *J. Anim. Sci.* 76(Suppl.1):185.

Barth, K. M., G. W. Vander Noot, W. S. MacGrath, and E. T. Kornegay. 1966. Nutritive value of garbage as a feed for swine. II. Mineral content and sup-plementation. *J. Anim. Sci.* 25:52-57.

Bath, D., J. Dunbar, J. King, S. Berry, and S. Olbrich. 1998. By-products and Unusual Feedstuffs. Feedstuffs® 1998 Reference Issue. Minnetonka: The Miller Publishing Co.

CAST (Council for Agricultural Science and Technology). 1995. Waste Management and Utilization in Food Production and Processing. Council for Agricultural Science and Technology. October, 1995. Ames, IA.

Clark, W. S. 1979. Our industry today: whey processing and utilization, major whey product markets. — 1976. *J. Dairy Sci.* 62:96-98.

Ensminger, M. E., J. E. Oldfield, and W. W. Heinemann. 1990. Feeds and Nutrition. 2d Ed. Clovis: The Ensminger Publishing Co..

Fontenot, J. P., G. J. Flick, and V. G. Allen. 1982. *Utilization of seafood waste as rumi-nant feed.* Annual Project Report. Sea Grant College, Virginia Polytechnic Institute and State University, Blacksburg.

Franklin and Associates. 1998. Characterization of Municipal Solid Waste in the United States: 1997 Update. U.S.Environmental Protection Agency. Municipal and Industrial Solid Waste Division. Office of Solid Waste. Report No. EPA530-R-98-007. Prairie Village, KS:Franklin Associates, Ltd.

14   *Westendorf*

Henry, W. A. and F. B. Morrison. 1920. Feeds and Feeding. 17th Ed. Madison: The Henry-Morrison Co.

Koch, A. R. 1964. The New Jersey swine industry - an economic analysis. New Jersey Agricultural Experiment Station Mimeo A. E. 297. Rutgers, New Brunswick, NJ.

Kornegay, E. T., G. W. Vander Noot, W. S. MacGrath, and K. M. Barth. 1968. Nutritive value of garbage as a feed for swine. III. Vitamin composition, digestibility, and nitrogen utilization of various types. J. Anim. Sci. 27:1345-1349.

Kornegay, E. T., G. W. Vander Noot, K. M. Barth, G. Graber, W. S. MacGrath, R. L. Gilbreath, and F. J. Bielk. 1970. Nutritive evaluation of garbage as a feed for swine. Bull. No. 829. College of Agric. Environmental Sci. New Jersey Agric. Exp. Sta. Rutgers, New Brunswick, NJ.

McClure, K. E., E. W. Klosterman, and R. R. Johnson. 1970. Feeding garbage to cattle and sheep. Ohio Report on Research and Development in Agriculture, Home Economics and Natural Resources. Volume 55:78-79. Ohio Agric. Res. Dev. Center, Wooster.

Minkler, F. C. 1914. Hog cholera and swine production. Circular No. 40. New Jersey Agricultural Experiment Station. Trenton, NJ.

Myer, R. O., J. H. Brendemuhl, and D. D. Johnson. 1999. Evaluation of dehydrated restaurant food waste products as feedstuffs for finishing pigs. J. Anim. Sci. 77:685.

Myer, R. O., T. A. DeBusk, J. H. Brendemuhl, and M. E. Rivas. 1994. Initial Assessment of Dehydrated Edible Restaurant Waste (DERW) as a Potential Feedstuff for Swine. Res. Rep. Al-1994-2. College of Agriculture. Florida Agricultural Experiment Station. University of Florida. Gainesville, FL.

NRC. 1983. Underutilized Resources as Animal Feedstuffs. National Research Council. Washington DC. National Academy Press.

NJAES. 1919. II. Garbage as a Hog Feed. Fortieth Annual Report of the New Jersey Agricultural Experiment Station. Trenton, NJ.

Pigott, G. M. 1981. Seafood waste management in the Northwest and Alaska. Report No. 40. In W. S. Otwell (Ed.). Seafood Waste Management in the 1980s: Conference Proceedings. Sea Grant College Program, University of Florida, Gainesville.

Prokop, W. H. 1996. The rendering industry - a commitment to public service. D. A. Franco and W. Swanson, eds. The Original Recyclers. National Renderers Association. Merrifield, VA.

Rivas, M. E., J. H. Brendemuhl, D. D. Johnson, and R. O. Myer. 1994. Digestibility by Swine and Microbiological Assessment of Dehydrated Edible Restaurant Waste. Res. Rep. Al-1994-3. College of Agriculture. Florida Agricultural Experiment Station. University of Florida. Gainesville, FL.

Tolan, A. 1983. Sources of food waste, UK and European aspects. Page 15-27. In: D. A. Ledward, A. J. Taylor, and R. A. Lawrie (Eds.). Upgrading Waste for Feed and Food. Butterworths, London.

U.S. Congress. 1980. Swine Health Protection Act. Public Law 96-468.

USDA. 1997. Agriculture Fact Book - 1997. USDA-Office of Communications. Washington, DC.

USDA-APHIS, VS 1990. Heat-Treating Food Waste—Equipment and Methods. USDA Animal and Plant Health Inspection Service, Veterinary Services. Program Aid No. 1324.

USDA-APHIS, VS. 1995. Risk Assessment of the Practice of Feeding Recycled Commodities to Domesticated Swine in the United States Department of Agriculture Animal and Plant Health Inspection Service, Veterinary Services. Centers for Epidemiology and Animal Health. Fort Collins, CO.

USDA-ERS. 1992. Animal Feeds Compendium. M. S. Ash, Ed. USDA-Economics Research Service. Agricultural Economic Report Number 656. PP. 65-110.

Walker, P. and T. Kelly. 1997. Selected fractionate composition and microbiological analysis of institutional food waste pre- and post-extrusion. In Proc. 2nd. Food Waste Recycling Symposium. (Westendorf and Zirkle, Ed.) New Jersey Department of Agriculture and Rutgers Cooperative Extension. Trenton and New Brunswick, NJ.

Waterman, J. J. 1975. Measures, stowage rates and yields of fishery products. *Torry advisory note no. 17.* Torry Research Station. Aberdeen, Scotland.

Westendorf, M. L., T. Schuler, and E. W. Zirkle. 1999. Nutritional quality of recycled food plate waste in diets fed to swine. Prof. Anim. Sci. 15(2):106-111.

Westendorf, M. L., Z. C. Dong, and P. A. Schoknecht. 1998. Recycled cafeteria food waste as a feed for swine: nutrient content, digestibility, growth, and meat quality. *J. Anim. Sci.* 76:3250.

Westendorf, M. L., E. W. Zirkle, and R. Gordon. 1996. Feeding food or table waste to livestock. *Prof. Anim. Sci.* 12(3):129-137.

Wohlt, J. E., J. Petro, G. M. J. Horton, R. L. Gilbreath, and S. M. Tweed. 1994. Composition, preservation, and use of sea clam viscera as a protein supplement for growing pigs. *J. Anim. Sci.* 72:546.

NOTE

1. Feedstuffs® 1998 Reference Issue. Address: Feedstuffs, 12400 Whitewater Dr., Suite 160, Minnetonka, MN 55343. 81997 The Miller Publishing Co.

2

Food Residuals: Waste Product, By-product, or Coproduct

by Paul Walker

Defining terminology can have far-reaching implications. The old adage, "what's in a name is everything," can have substantial impact regarding society's perception of a product. Therefore, the term(s) used to describe food waste can affect the status by which it is perceived. However, whether food waste is referred to as garbage, food residuals, edible residual material, plate waste, table waste, etc. may not be as important in the development of food-waste recycling for livestock feed as an industry, as whether or not the resulting product achieves coproduct status. If food-waste recycling (or what ever term ultimately describes this waste material) is to become a viable self-sustaining industry, the resulting product(s) must eventually achieve recognition as a coproduct. This recognition is as important as the food items generating the waste.

There is, generally, a recognized sequence of events or series of product development steps that a waste material must pass through to reach coproduct status. The sequence of descriptive terms that characterize this progression are, in order of occurrence: waste material, waste product, by-product, coproduct.

Generally, waste materials such as garbage have little, if any, value. The mentality for dealing with these materials centers on the most economical means for disposal. Traditionally, the most economical method for disposal of garbage has been landfilling. Without source separation of contaminants (in this case paper, plastic, metal, etc.) from the item destined for recycling (food waste), landfilling may remain the most economical alternative, even at higher relative costs. If perceived value can be realized from a waste selected for recycling, then waste material such as food waste can realistically progress through a series of events culminating in the achievement of coproduct status.

17

In order for garbage to achieve recognition as a waste product, i.e., food waste, several criteria must be met that include (1) an alternative use must be identified for the waste material, (2) the use must be accurately described, and (3) standards of quality must be established or at least identified.

The first of these three criteria has been achieved, primarily as a result of historical use. Garbage and food waste have been used as livestock feed for centuries with varying degrees of success. The second criterion has been well-documented for feeding garbage (unprocessed food waste) to swine. The role of processed food waste is less clearly defined. Ground, dehydrated, extruded, or otherwise processed food waste will probably be described as a feed additive for combining with other traditional and nontraditional feedstuffs, and eventual end-use in the diets of livestock and companion animals. In some situations, processed food waste may, depending on age of the animal, stage of the animal's production cycle, and species, serve as the sole dietary ingredient. Standards of quality for food waste have not been descriptive, at least for food waste in general. The contents, sources, and uses of food waste as a feedstuff vary so much that it may be impossible or impractical to establish uniform standards of quality for generic food waste. However, general descriptors of quality that all categories of food waste should adhere to may be identified, such as minimum nutrient densities, digestibilities, freedom from pathogens, absence of contaminants, etc. The primary reasons any identified food waste may remain a waste material are

- low energy or protein content
- high moisture content
- presence of contaminants
- lack of an AAFCO definition
- requirement for expensive processing
- high transportation cost
- lack of FDA approval

While feedstuffs are sources of vitamins and minerals, they are added to diets primarily for their energy and protein value. Dietary energy can be supplied by fat, carbohydrates, and protein when in excess. Most of the literature reports food waste to be relatively high in fat, 14 to 16% on a dry matter basis, and moderately high in protein, 20 to 28% on a dry matter basis (Barth et al. 1966; Flores et al. 1993; Heitman, et al. 1956; Kornegay et al. 1965; Kornegay et al. 1968; Kornegay et al. 1970). These data suggest food waste is moderate in energy and may have a greater contribution as a protein source in most diets. The source and type of food waste affects its use in balancing diets. Food waste may be an excellent source of vitamins depending on the method of processing. Some heat processing methods that utilize excessively high temperatures for extended periods of time may destroy some heat-labile vitamin activity. Food waste, also, is an excellent source of some minerals. Researchers (Myer et al. 1994; Walker and Wertz 1994; Walker et al. 1997) have expressed concern regarding relatively high sodium levels in food waste but reports to date (Myer et al. 1994; Walker et al. 1998) have observed sodium levels in food waste within acceptable ranges that can be accommodated in balanced diets. Walker et. al. (1997) evaluated pulped university cafeteria food waste for 13 selected elements (calcium, phosphorus, zinc, magnesium, manganese, cobalt, aluminum, chromium, nickel, potassium, copper, iron, and sodium) and found all element concentrations well below the dietary maximum tolerable limits (MTL) for beef, swine, and sheep. Accordingly, food waste appears to have sufficient nutrient density to allow progression from waste-material status to waste-product status.

A major concern of many nutritionists regarding food waste is its high moisture content. The dry matter content of food waste is as variable as food waste itself. Reported values range from 90% dry matter for cereals to 20% dry matter for some residential food waste. High moisture (anything more than 20%) food waste presents handling, storing, processing and feeding problems. The nutrient content of food waste is in the dry matter portion, not in the water fraction. High water content in food waste (more than 40% moisture) also may decrease animal consumption and reduce average daily dry matter intakes thereby reducing animal performance. Transportation costs are excessive if food waste contains high moisture content and may become a limiting factor for food-waste utilization. Moisture contents greater than 20% limit the length of time food waste can be stored (usually 1 to 2 days) prior to processing or feeding without excessive spoiling occurring. High moisture food waste can be stored for longer periods if it is refrigerated during warm weather or treated with enzyme/bacterial innoculants to allow the food waste to ferment. High moisture food waste can freeze during extreme cold weather, which limits usability unless precautionary measures are taken. High moisture content is problematic for most processing methods and increases the costs for processing food waste. In some instances, the energy expenditure to dry, dehydrate, or extrude food waste to remove excess water exceeds its value as a feedstuff. Economical methods for handling wet food wastes and for removing the water fraction must be identified and investigated if high moisture (greater than 20%) food waste progresses from waste-material status to waste-product status.

Another factor that can limit food waste's progression to by-product status is the presence of contaminants. Paper is a major contaminant in food waste, not recognized as an approved feed additive by the FDA. The presence of incidental paper contamination in food waste may be tolerated by the FDA if the processor is making a good-faith attempt to remove paper from the food waste. Mechanized technology of grinding and using air pressure to remove paper from dry food wastes such as cereals, crackers, etc. have been developed and used successfully. The most cost-effective method for separating paper from high moisture food wastes, such as cafeteria or restaurant food waste, is at the source of serving or consumption. Successful source-separation requires extensive and continuous public education. While practical, consumer education is time consuming.

Until 1999, generic food waste lacked an Association of American Feed Control Officials (AAFCO) definition. One purpose of AAFCO is to develop definitions and policies for animal feeds that can be used by state and federal regulatory agencies to monitor the effectiveness and usefulness of animal feeds. Until 1999, each individual food waste required an AAFCO definition. This requirement limited the use of generic food waste as a feedstuff. Since 1999, AAFCO has recognized at least two definitions for generic food waste, T60.96 Food-Processing Waste and T60.97 Restaurant Food Waste, providing definitive descriptions of food waste that facilitate its marketability as a by-product. These definitions also provide a basis for recognition of food waste as a feed additive by the FDA, which is essential if it is to receive universal acknowledgement in its progression from waste material to waste-product to by-product status.

Because of its variability, food waste must be processed to create a uniform product with consistent fractional composition. Variability in the dry matter, protein, fat, energy, and fiber content of food waste can limit its incorporation as a feed additive into livestock and companion animal feeds. Processing (grinding, drying, blending, etc.) of food waste improves its marketability. Commercial scale grinders, dehydrators, extruders, and dryers are expensive. To ensure economical processing costs, economy of scale becomes increasingly important. The greater the volume of a product processed daily through one manufacturing plant, the more competitively priced food waste becomes as a feed additive. The speed with which food waste processors are established will impact the rate at which food waste moves from classification as a waste material to that as a waste product.

Perhaps the greatest factor limiting food waste's attainment of by-product status is the collection and transportation of the raw

Table 2.1. Characteristics required for obtaining by-product status

Nutrient dense
High digestibility
Low water content
Absence of contaminants
Capable of prolonged storage
Minimal processing required
Nonprohibitive transportation and handling
Cost effective
High public acceptance
Legal

unprocessed material from its source of generation to a processing facility. The higher the relative water content, the higher the dry matter transportation cost. As long as alternative disposal options (landfilling, for example) remain low priced, progression of food waste to by-product status will be slow. As collection costs and landfill tipping fees increase, food waste's recognition as a waste-product will increase proportionately. There are 10 characteristics that food waste as a waste product must achieve prior to progressing to by-product status (table 2.1). If any one of these characteristics is lacking or is not successfully addressed, food waste will not be recognized as a by-product.

Food waste previously has been cited as a feedstuff of varying yet considerable nutrient density (Flores et al. 1993; Kornegay et al. 1965; Walker and Wertz 1994; Walker et al. 1997). Diets containing Okara (a by-product of soybeans processed for tofu production), granola bars, bakery waste (bread and cookie discards from supermarkets), and brewers grains (by-product of the brewing industry) were readily consumed by ducks (Farhat et al. 1998); however, birds fed the diets amended with these food wastes consumed higher proportions of fat and less protein than controls. Chemical analyses of dehydrated edible restaurant waste have reported this waste stream as moderately high in protein, high in fat and relatively low in crude fiber and ash (Kornegay et al. 1970). Proximate analysis of fresh pulped food waste generated by university dining centers was found to contain relatively high levels of crude protein ($29.4 \pm 7.2\%$) and fat ($15.8 \pm 3.2\%$). These independent analyses of food waste are typical of the ranges of reported fractionate values for food waste. An often recognized assumption is that food wastes, in general, contain substantial amounts of protein and energy. However, classification of food waste as primarily a protein feedstuff, energy feedstuff, or mineral source, etc. is difficult, if not impossible, because of the variable composition of the waste stream sources generating a specific food waste. While this reported variability in composition may limit the broad classification of food waste as a particular "type" of feed additive, it does not prevent food waste from being recognized as a by-product. Some food waste such as discards generated from cereal and bakery manufacturing, have been established as economical energy substitutes in livestock and companion animal feeds. When the niche markets for the many food wastes are considered collectively, food waste is well on its way toward recognition as a by-product.

Less is known about the digestibility of food waste than is known about food waste composition. A general assumption, however, is that the digestibility coefficient for the primary nutrients found in large amounts in food waste must be 75% or greater if food waste is to become recognized as an economical feed additive for inclusion in traditional diets.

Some food-waste streams are high in dry matter (85%). Others are relatively low in dry matter, containing 30% or more water. The high moisture content of some food-waste streams will limit their usefulness as feed additives unless economical methods for removing the moisture fraction are identified. Because the energy charges associated with drying and dehydration are often cost restrictive, numerous food-waste processors blend wet food waste with other dry feedstuffs to lower the average moisture content prior to processing. This method of lowering the moisture content has proven effective for such processing technologies as extrusion. High transportation costs of wet food waste and limited storage times can limit some food waste from achieving by-product status.

Freedom from unwanted contaminants such as paper, plastic, etc. and elimination of potential pathogens are required characteristics for any by-product. Current technology exists that can ensure limited contamination of unwanted materials and reduced potential for pathogen content. Little reason exists why these technologies (grinding and forced air separation, extruding, dehydrating, etc.) can not be universally adapted for processing food waste, except that they may in some cases be cost prohibitive. Simple procedures, such as source separation of food waste from unwanted items (wrappings, shipping containers, etc.) at the site of generation and physical separation of food waste from other waste stream components on-site at a processing facility, may require substantial manual labor and, therefore, can be cost prohibitive. Mechanizing and automating these operations are the key elements in preventing contamination from being a limiting factor in the establishment of food waste as a by-product.

Production of dry product (dry matter equal to or greater than 88%) capable of prolonged storage in either bulk or bagged form is a characteristic common to most by-product feedstuffs. An exception, perhaps, is wet corn gluten, a by-product of the corn milling industry. Wet corn gluten feed (WCGF), with 50 to 60% moisture, has limited uses for feeding cattle because of its high water content. WCGF must be fed within 3 to 4 days following arrival at a cattle-feeding facility or substantial spoilage and nutrient loss will occur. Consequently, WCGF's utilization as a feedstuff is limited to large operations in geographic locations near wet milling plants. By-product feed additives capable of being utilized in a variety of animal feeds contain at least 88% dry matter. Dehydration, drying, and extrusion are processing methods that should be considered as part of most food waste processing plans if by-product recognition of the food waste is an objective. To be effective, any food-waste processing regimen should involve minimal processing and nonprohibitive transportation and handling.

Currently, many food-waste processing facilities are indirectly subsidized. Operators of food-waste processing plants often receive compensation for accepting the waste stream similar to or approaching the value of landfill tipping fees for garbage. Processors often charge generators of food waste a pick-up or reception fee. This fee offsets processing costs that are sometimes higher than the nutritive value of the feed additive produced. The subsidizing of food waste, processing costs with tipping fees, is not a sustainable practice and potentially limits food waste's ability to achieve recognition as a by-product.

In order to achieve by-product status, food waste's inclusion in animal feeds must be approved by federal and state regulations. For the purposes of this discussion, garbage can be defined as "meat-free" or "contains meat." There is less concern regarding the wholesomeness of vegetarian food waste than there is for food waste that contains animal protein. Federal laws pertaining to the feeding of food waste were originally put in place to prevent the introduction and/or spread of certain diseases into the swine industry of the United States. These diseases include foot-and-mouth disease, African swine fever, hog cholera, vesicular exanthema, *Trichinella spiralis*, San Miguel sea lion virus, *Salmonella*, tuberculosis, pseudorabies, and erysipelas. These diseases are transmitted through contact with raw muscle tissue (meat) from infected animals. Therefore, in order to prevent the transmission of these diseases, federal regulations require that all garbage containing meat be boiled for 30 min at 212° F (100° C) before being fed to livestock. Meat as a component of garbage is described explicitly in the federal definition for food material classified as garbage.

GARBAGE. All waste material derived in whole or in part from the meat of any animal (including fish and poultry) or other animal material, and other refuse of any character whatsoever that has been associated with any such material, resulting from the handling, preparation, cooking, or consumption of food, except that such term shall not include waste from ordinary household operations which is fed directly to swine on the same premises where such household is located. (1990 Code of Federal Regulations, Title 9, Chapter 1, Subchapter K, Part 166.1)

In other words, if a food waste contains any meat, it is defined as garbage under the federal definitions. Thus, the garbage must be "treated" (boiled for 30 min.) before it can be used as an animal feed. The federal definition of treated garbage is

TREATED GARBAGE. Edible waste for animal consumption derived from garbage (as defined in this section) that has been heated throughout at boiling or equivalent temperature (212° F or 100° C at sea level) for 30 (thirty) minutes under the supervision of a licensee. Part 166. I.

The federal definitions of garbage and treated garbage are then incorporated into the restrictions of Part 166.2.(a):

No person shall feed or permit the feeding of garbage to swine unless the garbage is treated to kill disease organisms, pursuant to this Part, at a facility operated by a person holding a valid license for the treatment of garbage; except that the treatment and license requirements shall not apply to the feeding or the permitting of the feeding to swine of garbage only because the garbage consists of any of the following: rendered products; bakery waste; candy waste; eggs; domestic dairy products (including milk); fish from the Atlantic Ocean within 200 miles of the continental United States or Canada; or fish from inland waters of the United States or Canada which do not flow into the Pacific Ocean.

The above federal restrictions exempt from treatment include bakery waste, candy waste, eggs, domestic dairy products, and fish from the Atlantic Ocean. (Fish from the Atlantic Ocean are exempt because the San Miguel sea lion virus has only been linked to sea lions found in the Pacific Ocean.) This is because these nonmeat food by-products by themselves are not carriers of the diseases mentioned previously. The diseases are carried in the muscle tissue of infected animals. As long as the above nonmeat food by-products do not come in contact with infected muscle tissue, the viral or bacterial disease-causing organisms cannot be transmitted.

Regulations for feeding food waste vary greatly by individual states because each state retains the right to establish its own laws even when federal guidelines exist. In this particular regulatory situation, states have the option of passing laws that are more, or equally as restrictive, as the federal laws. In fact, a wide variation in state laws and definitions can be observed on a state-by-state breakdown. In each case, how a state defines the term garbage determines what food materials, if any, can be fed to livestock. Some states (i.e., Georgia, Iowa, Nebraska, New York, and Wisconsin) do not allow any food waste (according to their definition of garbage) to be fed to any animal, while other states (i.e., California, Nevada, New Jersey, and North Carolina) have guidelines that closely resemble the federal law. In the fall of 1997, the FDA adopted the Mammalian Protein-Ruminant Feed Ban, which is aimed at preventing bovine spongiform encephalopathy. This ban excludes feeding ruminants food waste containing animal protein that has not been offered for human consumption nor heat processed. Consequently, it is now illegal to feed food waste containing animal protein to cattle and sheep. The ban does, however, include the following exemption "inspected meat products which have been cooked and offered for human food and further heat processed for feed (such as plate waste and used cellulosic food casing) (Fed. Reg. 30976.)" The phrase "further heat processed" may include cooking at 212° C for 30 min, dehydration, or extrusion.

Even though the feeding of food waste to animals may be legal, sufficient volumes of food waste will not be recycled as animal feed if society has a negative perception of feeding food waste. Low public acceptance of food waste as an animal feed ingredient will prevent recognition of food waste as a by-product. To date, popular press and public opinion influencers have readily accepted the concept of recycling food waste as animal feed. Food waste recycling is viewed as a "green idea" and feeding human food waste to animals is gaining favorable attention in both developed and underdeveloped countries.

The ultimate recognition any waste material can achieve is that of coproduct status. Coproduct status infers that the by-product is as valuable or has as much demand as the original product from which it was derived. Coproduct recognition implies that the product has a standardized price, uniform composition, and a commercial identity regarding its function or potential uses. How rapid food waste attains coproduct status is dependent on the length of time required for food waste to be recognized as a by-product. Coproduct recognition is driven by increased demand for a by-product.

A Historical Perspective of Soybean Meal

The classical success story of a waste obtaining co-product status is the history of soybean meal. The soybean came to America in the 18th century to the Atlantic Coast as ballast in sailing vessels. The first published account of the soybean plant in the U.S. appeared in 1804. It was introduced to the United States from China as an oddity in a garden in Pennsylvania. In 1850, the U.S. Department of Agriculture (USDA) started research to study the use of the crop as forage, green manure, silage or hay. Even up until 1940, there were more acres of soybeans harvested for hay than for beans in the United States.

The first domestically produced soybeans were processed for oil in 1915 in a North Carolina cotton seed mill, though an imported lot of soybeans were processed in Seattle, Washington in 1911 when Pacific Oil Mill brought the soybeans from Manchuria. World War I caused a shortage of edible vegetable oils and the versatile bean filled the gap. The soybean processing industry expanded and the USDA soon encouraged farmers to produce more soybeans for oil, rather than forage. In 1935 when restrictions on corn acreage began, soybeans were one of the alternate crops. World War II again increased demand for edible vegetable oils. As a result of processing soybeans for their oil content mounds of soybean meal were being generated as a waste with no apparent use. Through some innovative applications soybean meal was found to have value as a protein supplement in livestock rations. However, growth depressions were sometimes observed when soybean meal was added to non-ruminant diets. Discovery of the anti-nutritional factors present in raw soybeans and their control through adequate heat processing aided in recognizing the true value of soybean meal. Scientific advances demonstrated that soybean meal rations could be fortified with vitamins, especially B_{12} and the limiting amino acid Methionine, thus eliminating the need of supplemental animal protein. Soybean meal had become a waste product.

Originally soy oil was extracted by means of hydraulic or screw press methods which left 3-6% oil in the meal. As more demand for vegetable oil was demonstrated, the solvent extraction method was adapted to remove the majority of the soy oil and use of the meal as a livestock feedstuff increased. Soybean meal was now considered as a by-product.

Today poultry consume about 50% and swine over 30% of all domestically produced soybean meal. The other 20% is consumed by beef and dairy cattle, sheep, fish and humans or is used in industrial applications. Futures contracts for soybean meal are bought and sold on the Chicago Board of Trade and soybean meal is recognized as a commodity of equal status to that of soy oil. Soybean meal is a co-product produced from soybeans - the ultimate success story for a material that was once considered a burdensome waste. (This history of soybean meal was written in consultation with Nabil Said, Director of Technical Services for Triple "F" Inc. Insta-Pro®, Des Moines, Iowa.)

Some sources of food waste lend themselves to producing food waste capable of becoming a coproduct better than others. Sources of food waste can be ranked from high quality to low quality. An order frequently recognized among food waste recyclers from highest to lowest is institutional, luxury hotels, upscale restaurants, fast food establishments, and households.

Institutions such as university dining centers, retirement home cafeterias, prisons, and hospitals generate food waste that is nutrient dense (high in fat and protein), contains minimum contamination (contains little paper, plastic, etc.), and is consistent in composition as determined by a fractionate analysis from day to day. Luxury hotel dining rooms and upscale restaurants generate food waste of greater nutritional value than fast-food establishments. The waste stream generated by fast-food restaurants contains a higher proportion of paper, plastic, and styrofoam than it does food. Household garbage will contain greater contamination of unwanted food items than the other sources of food waste mentioned. Accordingly, recycling of food waste as animal feed will be driven by those who capitalize on receiving the food waste generated by institutions and upscale hotels and restaurants.

There are three primary motivations that drive the generators of food waste and others to consider recycling food waste as a feed additive. These motivations are entrepreneurship, federal and state mandates, and environmental stewardship. If substantial money can be realized from recycling food waste as a livestock feedstuff, then individuals and companies within the business community will become involved. Several companies have already been established and are enjoying considerable financial success processing dry (90% or greater dry matter), discarded cereal, and bakery products into animal feed additives. A greater challenge exists for processing high moisture (greater than 20% water) food waste into animal feed. Operations that can generate revenue collecting and transporting wet food wastes, processing mixed ingredient food wastes into a dry uniform end product, and locating multiple end users for the processed food waste will be innovative users of existing and new technology. Research conducted at several universities, and by several industrial manufacturers have laid a sound foundation for venture capital to commercialize the processing of wet food waste into animal feed.

State and federal mandates have been purposefully adopted to indirectly drive both private and public entities to recycle. Legislation regulating solid waste recycling has been approved at both the state and federal level. Many mandated and several voluntary programs have had positive effects on encouraging innovative recycling initiatives. Regulation in several states banning landscape waste from landfill disposal is one example of mandates that have reduced the solid waste stream. Higher tipping fees resulting from reduced landfill capacity are encouraging investigation of food waste recycling opportunities on the east and west coasts of the United States. Innovative food waste recycling initiatives have occurred infrequently in the midwest, in large part, because landfill tipping fees are relatively inexpensive compared to alternative opportunity costs. In addition, few midwestern states have adopted any legislation regulating food waste disposal. Several states do, however, regulate or prohibit the feeding of garbage (food waste) to livestock and follow the policies stated in the federal Swine Health Protection Act.

Many innovative recycling initiatives occur because people believe in environmental stewardship. These "do-gooders" who "wear a white hat" are driven less by a desire to make a profit and more by their moral beliefs. These individuals are key components who will play a major role in developing innovative food-waste recycling programs. The greatest value for recycling food waste as animal feed and creating coproduct status for food waste may be the net return to society. The net economic return to livestock feeders and feed processing companies may be of marginal value relative to the improved health, perceived or real, of our environment. Solid waste disposal savings [food waste recycled (tons) times the landfill tipping fee (dollars:ton)] and feed savings (nutritional replacement value) for food waste may be the means that determine if food waste becomes a coproduct, but neither may be as important as the improved environmental health enjoyed by a society participating in food-waste recycling.

References

Barth, K. M., G. W. Vander Noot, W. S. MacGrath, and E. T. Kornegay. 1966. Nutritive value of garbage as a feed for swine. II. Mineral content and supplementation. *J. Anim. Sci.* 25:52.

Farhat, A., L. Normand, E. R. Chavez, S. P. Touchburn, and P. C. Lague. 1998. Comparison of growth performance, carcass yield and composition and fatty acid profiles of Pekin and Muscovy ducklings fed diets based on food wastes. Unpublished data. Dept. Anim. Sci. McGill Univ. Ste. Anne de Bellevue Quebec, Canada.

Flores, R. A., D. A. Ferris, M. K. King, and C. W. S. Shanklin. 1993. Characterization of food waste streams: a proximate analysis of plate and production wastes from university and military dining centers. American Society Agricultural Engineering Annual Meeting. Chicago, IL.

Heitman, H., Jr., C. A. Perry, and L. K. Gamboa. 1956. Swine feeding experiments with cooked garbage. *J. Anim. Sci.* 15:1072.

Kornegay, E. T., G. W. Vander Noot, W. S. MacGrath, J.G. Welch, and E.D. Purkhiser. 1965. Nutritive value of garbage as a feed for swine. I. Chemical composition, digestibility and nitrogen utilization of various types of garbage. *J. Anim. Sci.* 24:319.

Kornegay, E. T., G. W. Vander Noot, W. S. MacGrath, and K. M. Barth. 1968. Nutritive value of garbage as a feed for swine. III. Vitamin composition, digestiblity and nitrogen utilization of various types. *J. Anim. Sci.* 27:1345.

Kornegay, E. T., G. W. Vander Noot, K. M. Barth, G. Garber, W. S. MacGrath, R. L. Gilbreath, and F. J. Bielk. 1970. Nutritive Evaluation of Garbage as a Feed for Swine. Bull. No. 829. College of Agriculture and Environmental Science. New Jersey Agricultural Experiment Station. Rutgers, New Brunswick, NJ.

Myer, R. O., T. A. DeBusk, J. H. Brendemuhl, and M. E. Rivas. 1994. Initial Assessment of Dehydrated Edible Restaurant Waste (DERW) as a Potential Feedstuff for Swine. Res. Rep. A1-1994-2. College of Agriculture. Florida Agricultural Experiment Station. University of Florida. Gainesville, FL.

Walker, P. M. and A. E. Wertz. 1994. Analysis of selected fractionates of a pulped food waste and dish water slurry combination collected from university cafeterias. Abst. *J. Anim. Sci.* 72:523. Suppl. 1.

Walker, P. M., S. A. Wertz, and T. J. Marten. 1997. Selected fractionate composition and digestibility of an extruded diet containing food waste fed to sheep. Abst. *J. Anim. Sci.* 75:253. Suppl. 1.

Walker, P. M., F. B. Hoelting, and A. E. Wertz. 1998. Fresh pulped food waste replaces supplemental protein and a portion of the dietary energy in total mixed rations for beef cows. *The Prof. Anim. Scien.* 14:207-16.

3

Regulation of Food-waste Feeding: The Federal Perspective

by Daniel G. McChesney

Legal Authority

The FDA has primary responsibility in the federal government for food safety in the United States. The FDA is charged with the enforcement of the Federal Food, Drug, and Cosmetic Act (FFDCA or the Act) (FFDCA 1998)and related food safety aspects of the Public Health Service Act (PHSA 42 U.S.C.). The FDA mandate under the PHSA and the FFDCA includes widespread responsibilities to help ensure preharvest food safety. For example, one mission of the FDA's Center for Veterinary Medicine (CVM) is to regulate the levels of contaminants permitted in animal feeds to ensure that the food for man and animals is safe and free of illegal drugs, industrial chemicals, pesticide residues, and harmful bacteria. The USDA has responsibility for the safety of human food products resulting from the slaughter of most food animals.

The FFDCA (§201(f))defines food as "articles used for food or drink for man or other animals . . . and articles used for components of any such article." Therefore, any waste product, garbage, by-product, or coproduct, regardless of source, that is intended to be used as a feed ingredient or to become part of an ingredient or feed, is considered a "food" under the FFDCA and thus subject to regulation by the FDA. Furthermore, it is the position of the FDA that a product intended for use as a feed or feed ingredient must not be adulterated as defined in Section 402(a) of the FFDCA. Section 402(a) of the Act has numerous provisions for establishing adulteration. The most appropriate subsections of 402 to apply to garbage, waste products, by-products, and coproducts are (a)(1) and (a)(3). Section 402(a)(1) states in part that a food (feed) shall be deemed to be adulterated "if it bears or contains any poisonous or deleterious substance which may render it injurious to health" and subsection (a)(3) states in part "a food shall be deemed to be adulterated if it is otherwise unfit for food (feed)." Additionally, Section 402(a)(2)(C) states that a food (feed or feed ingredient) can be considered adulterated if "it bears or contains any food additive which is unsafe (unapproved) within the meaning of Section 409."

The legal status of whether a product should be considered Generally Recognized As Safe (GRAS) or an unapproved food additive is often questioned and debated. The decision on this status is largely based on whether there are significant safety concerns related to the product or a similar product, whether there is currently an approved food additive use for the product, and its history of use in animal feed. The pivotal issue in the decision is whether there is sufficient safety data available in the open scientific literature that would enable an unbiased panel of experts to judge the safety of the product. If such data exist, the ingredient is a good candidate for being considered GRAS and allowable in animal feed via the Association of American Feed Control Officials (AAFCO) definition process. If the data are not available or the experts disagree on the interpretation, then the ingredient very likely will have to undergo the food additive process.

If a substance was used in food before 1958, general recognition that the use is safe can be based on scientific procedures or experience based on common use in food (62 FR 552-566; §201(s) of the Act (21 U.S.C. 321(s)); and 21 CFR 570.30(a)). General recognition of safety, through experience based on common use in food, prior to January 1, 1958, may be determined without the quantity or quality of scientific procedures required for approval of a food additive regulation, but it nonetheless requires a demonstration of (1) safe use based on common use and (2) an expert consensus of safety, based on that common use (21 CFR 570.30). The simple assertion of this safe use thus does not satisfy the burden the proponents of the use bear to establish general recognition.

Moreover, even if a substance is GRAS based on common use in food or on scientific procedures, the FDA may reassess the GRAS status of a food ingredient based on new information (21 CFR 530.30(g); see also, e.g., 51 FR 25021, July 9, 1986 (Sulfiting Agents; Revocation of GRAS Status for Use on Fruits and Vegetables to be Served or Sold Raw to Consumers)). Thus, even if an ingredient of a feed was GRAS based on common use in feed prior to 1958, that does not preclude the FDA from reassessing it now that new studies, data, or other information exist that show that the substance is, or may be, no longer safe (this is true whether the studies or data are published or unpublished (50 FR 27294-27296 (July 2, 1985)) or that there is no longer the basis for an expert consensus that it is safe.

Expert opinion that the substance is GRAS would need to be supported by scientific literature and other sources of data and information. General recognition cannot be based on an absence of studies that demonstrate a substance is unsafe; there must be studies or other information to establish that the substance is safe (see U.S. v. An Article of Food * * * Coco Rico, 752 F.2d 11 (1st Cir. 1985)). Furthermore, if there are studies and other data or information that raise questions about the safety of the use of the material, this conflict—just like a conflict in expert opinion—may prevent general recognition of the substance. This conflict in expert opinion can result in an ingredient no longer being categorically regarded as safe (62 FR 552-566).

Because the expert opinion must be general, a substance is not GRAS if there is no recognition among experts or there is a genuine dispute among the experts as to whether it is safe. Although there need not be unanimity among qualified experts, that a substance is safe for "general recognition" of its safety to exist, an "expert consensus" is required (see Weinberger v. Hynson, Wescott & Dunning, Inc., 412 U.S. 606, 632 (1073)). When there is a dispute among experts as to general recognition, The * * * issue (of actual safety) is to be determined by the FDA which, as distinguished from a court, possesses superior expertise usually of a complex scientific nature for resolving that issue (United States v. 50 Boxes).

As part of the FDA's commitment to achieving the goals for the Reinventing Food Regulations section of the President's National Performance Review, the agency undertook a review of the procedures by which a substance can receive GRAS status. Based on this review, the FDA proposed to clarify the criteria for exempting the use of a substance in human food or in animal feed from the premarket approval requirements of the FFDCA because such use is generally recognized as safe (GRAS). The FDA also proposed to replace the current GRAS affirmation process with a notification procedure whereby any person may notify FDA of a determination that a particular use of a substance is GRAS. Under the proposed notification procedure, the notice would include a "GRAS exemption claim," dated and signed by the notifier, that would provide specific information about a GRAS determination in a consistent format. This claim would include a succinct description of the notified substance, the applicable conditions of use, and the basis for the GRAS determination. The GRAS exemption claim would also include a statement that the information supporting the GRAS determination was available for FDA review and copying or would be sent to the FDA upon request. In addition to the GRAS exemption claim, the notice would include detailed information about the identity of the notified substance and a detailed discussion of the basis for the notifier's GRAS determination.

The FDA would evaluate whether the notice provided a sufficient basis for a GRAS determination and whether information in the notice, or otherwise available to the FDA, raised issues that would lead the agency to question whether use of the substance was GRAS. Within 90 days of the date of the notice's receipt, the FDA would respond to the notifier in writing and could advise the notifier that no problems were found with the notification or that the agency had identified a problem with the notification.

For each notice received, the FDA would make the GRAS exemption claim and the agency's response readily accessible to the public. While the FDA would maintain a readily accessible inventory of notices received and the agency's response to them, this inventory neither would be codified nor referenced in the agency's regulations.

As of the writing of this chapter (1999), the proposed regulation, which was published in the April 17, 1997, Federal Register, has not yet been finalized, and some of the provisions described above could change in response to comments to the proposed rule.

The Act defines a food additive as "any substance the intended use of which results or may reasonably be expected to result, directly or indirectly, in its becoming a component or otherwise affecting the characteristics of any food * * * if such substance is not generally recognized, among experts qualified by scientific training and experience to evaluate its safety, as having been adequately shown through scientific procedures (or, in the case of a substance used in food prior to January 1, 1958, through either scientific procedures or experience based on common use in food) to be safe under the conditions of its intended use * * *" (see section 201(s) of the act (21 U.S.C. 321(s))).

The definition of food additive in section 201(s) of the Act does not apply to substances used in accordance with a sanction or approval granted prior to enactment of section 201(s) of the Act and granted under the Act, the Poultry Products Inspection Act (21 U.S.C. 451 et seq.), or the Federal Meat Inspection Act (21 U.S.C. 601 et seq.).

Therefore, there are two approaches by which a previously unapproved nondrug product can be accepted for use in animal feed. The approaches are the Food Additive Petition (FAP) process administered by the FDA or the AAFCO (1999) definition process. The AAFCO definition process may be used for products with minimal or no safety concerns but have not met all of the requirements to be considered GRAS. Both processes require data on the safety, utility, and manufacturing process. Both also require information on the proposed use, species for which the use is intended, the amount to be used, and a proposed label.

The FAP process is an in-depth review of the human safety, animal safety, utility, and manufacturing of the compound. An environmental assessment was required for food additives in the past. However, many food additives now qualify for a categorical exemption based on guidelines for environmental assessments published in the Federal Register (July 29, 1997; vol. 62, no. 145, pages 40570-600) and entitled "National Environmental Policy Act; Revision of Policies and Procedure; Final Rule." The FAP process requires a substantial amount of data to be generated by the sponsor or gathered from the open scientific literature. If all the information is found acceptable, then the compound receives a formal approval and is listed in the Code of Federal Regulations. Revoking the approval is also a formal process and could require substantially more data then would be required to remove regulatory discretion.

The regulations regarding food additives and food additive petitions are located in 21 CFR 570 and 21 CFR 571, respectively.

The AAFCO definition process is reserved for compounds with no safety concerns and for which we are willing to apply regulatory discretion. Compounds in this category, like GRAS compounds (21 CFR §582.1), can be reassessed by the FDA as new information becomes available. If this information shows that the substance is, or may be, no longer safe or that there is no longer the basis for an expert consensus that it is safe, then regulatory discretion can be withdrawn at anytime.

BSE Regulation and Food Waste

The FDA amended its regulations to provide that protein derived from mammalian tissues for use in ruminant feed is a food additive subject to certain provisions in the FFDCA (1998). The final rule restricts the use of protein derived from mammalian tissues, with certain exceptions, in ruminant feed. The regulation also established a flexible system of controls designed to ensure that ruminant feed does not contain animal protein derived from mammalian tissues and to encourage innovation in such controls.

The agency has carefully considered the various exclusions and defined "protein derived from mammalian tissue" as any protein-containing portion of mammalian animals, excluding blood and blood products, gelatin, inspected, and processed meat products that have been cooked and offered for human consumption and further heat processed for feed (such as plate waste and used cellulosic food casings), milk products, and products whose only mammalian protein consists entirely of porcine or equine protein (21 CFR §589.2000).

The FDA excluded these items from the definition because the agency believes that they represent a minimal risk of transmitting transmissible spongiform encephalopathies (TSEs) to ruminants through feed (62FR 552-566; 62 FR 30936). The excluded proteins and other items are materials that available data suggest do not transmit the TSE agent, or have been inspected by the FSIS or an equivalent state agency at one time and cooked and offered for human food and further heat processed for feed. Thus, they are of lower risk than those products that the agency has determined to be non-GRAS.

The FDA propagated the regulation because ruminants could be fed protein derived from tissues in which TSEs have been found and such proteins may cause TSEs in ruminants. TSEs are progressively degenerative central nervous system diseases of man and other animals that are fatal. Epidemiologic evidence gathered in the United Kingdom suggests an association between an outbreak of a ruminant TSE, specifically bovine spongiform encephalopathy (BSE), and the feeding to cattle of protein derived from sheep infected with scrapie, another TSE. There is also an epidemiologic association between BSE and a form of human TSE known as new variant Creutzfeldt-Jakob disease (nv-CJD) reported in Europe. Neither BSE nor nv-CJD has been diagnosed in the United States, and the BSE regulation is intended to prevent the establishment and amplification of BSE in the United States through feed and thereby minimize any risk to animals and humans.

Whether plate waste should be excluded from the BSE regulation was the subject of several comments. The majority of the comments supported the exclusion of plate waste from the definition of protein derived from mammalian tissues. The comments explained that all food products that compose plate waste have already been cooked and inspected several times before being offered for human consumption and later thrown away. Commercial processors of plate waste dehydrate the product at temperatures reaching 290 to 400° F when converting it to an animal feed ingredient. The comments also asserted that the plate waste comes from institutions (universities, retirement homes, hospitals, prisons, etc.), fast-food establishments, and large restaurants and cafeterias and does not consist of tissues that have demonstrated infectivity in cattle, e.g., brain, spinal column, eye, and distal ileum of cattle. Other comments stated that plate waste consists mostly (approximately 98%) of nonmeat products and is high in moisture. The high moisture content requires the addition of 50 to 60% corn, soybeans, or similar products to aid in the dehydration and extrusion process. Also noted was that the feeding of plate waste remains a common practice in many parts of the United States and around the world and that plate waste comprises approximately 8.9% of the Municipal Solid Waste stream in the United States.

One comment, from the USDA/APHIS, opposed an exclusion for plate waste, stating that the exclusion was too broad and could be interpreted to be similar to the USDA definition for garbage at 9 CFR 166.1 and that trimmings (bone and nervous tissue) from TSE-susceptible species might be included under the exclusion.

The FDA agreed with the USDA/APHIS that the inclusion of trimmings or high-risk tissue, such as brain and eyes, is inappropriate for use in ruminant feed. The FDA's approach to eliminating trimmings was to describe an acceptable product as one that was cooked and offered for human consumption. This phrase satisfactorily addressed USDA/APHIS's concern and clarified the FDA's position with regard to raw meat products.

The FDA also declined to expand the exclusion to include all ruminant meat that has passed federal or state inspection for human consumption because this would have required the FDA to remove the safeguard against trimmings, alluded to above, and also would allow brains and eyes that have passed inspection, and are known to be high-risk material for the BSE agent, to be fed to ruminants. The agency acknowledged in the BSE regulation that accurately describing products acceptable under this exclusion is difficult. In general, the FDA interprets this exclusion as being restricted to food prepared in restaurants or restaurant-like establishments, offered to consumers for consumption on the premises, and then discarded by the consumer. Precooked food items, such as hot dogs, casings from cooked hot dogs, and cooked deli items, would be excluded from regulation by this exclusion.

In summary, the decision to exclude plate waste was based on the fact that a small proportion of meat is included in plate waste and that plate waste represents a small proportion of ruminant feed. Additionally, the heat and pressure used to process plate waste should further reduce the risk of transmitting the TSE agent through feed in a product that is of minimal risk prior to the processing as plate waste.

Enforcement

In the FDA's overall approach to enforcement, education plays a role. Regulation can impact industries that do not often see the federal government and/or that are not part of any trade organization. It is difficult for the regulated industry to comply when it does not know and/or understand the requirements. In the past, the FDA has tended to take the position that if you are part of the regulated industry, it is your responsibility to know the laws and regulations that apply. We now spend more effort involving the regulated industry in the development of regulations and policy and on education of the industry and affected parties to make sure they understand the requirements and how to meet them. The development of the BSE regulation is an example of this new approach. Industry was involved early in the process and a level of understanding and cooperation was established that would not be likely if the process began after the regulation was final.

Reasons for noncompliance can be many, but generally can be categorized as genuine mistakes or misunderstandings because a firm or individual has not received word about a regulation, intentional noncompliance through failure to correct problems noted during inspections, or intentional disregard for a regulation.

In the case of a genuine mistake or misunderstanding, the preferred course of action is education and reinspection provided an immediate safety issue is not involved. If an immediate safety hazard is involved, the FDA or the state would take action to remove the product from the market.

In the case of intentional noncompliance, the first action of choice would ordinarily be a warning letter. This letter notifies the responsible parties of a violation or violations and asks for a response within a certain time frame explaining corrective actions taken. When it is consistent with the public protection responsibilities of the FDA, and depending on the nature of the violation, it is common practice to afford individuals and firms an opportunity to voluntarily take appropriate and prompt corrective action. The warning letter is issued for the purpose of achieving this voluntary compliance and for prior notice of violations because there is an expectation that a majority of individuals and firms will voluntarily comply with the law. Warning letters are informal and advisory, communicating our position on a matter, but are not considered a final agency action. The agency does have additional, more stringent enforcement tools available that include product seizure, injunction, and prosecution when the warning letter is not effective or the noncompliance is egregious. Again, if an immediate safety hazard is involved, the FDA or the state would take action to remove the product from the market.

Finally, compliance with federal law and regulations generally represent the minimum requirements. State and local laws and regulations that are at least as stringent as federal requirements will almost always take precedent over the federal ones. Thus, compliance of a product with federal law is a requirement for interstate commerce, but it is not a guarantee that the product can legally enter commerce at the state or local level. Feeding garbage to swine is permitted under the Swine Health Protection Act (a federal law) (9 CFR 166), yet almost half of states do not permit the feeding of garbage to swine (9 CFR 166).

The Future

The definition of food has important implications for companies or municipalities wishing to market food waste as an animal feed or feed ingredient. These entities must realize that they are producing a food product and have an obligation to produce one that is safe and whole-some. In order to do this, they must consider the source, the ingredients, and the quality of the ingredients used in the principal product.

The potential resource conservation and economic benefits for the use of nontraditional sources of animal feed ingredients and feed can be substantial. However, with these benefits also comes the potential for introducing contaminants into the feed supply and thus indirectly into the human food supply. The variety of waste products, by-products, and coproducts being considered for use in animal feed is growing rapidly and improvements in processing technology and their applications are making the use of these products more economically viable. In the past, federal and state governments have established regulatory guidance (tolerances, action levels) for safety-related issues, and the AAFCO has established definitions for product identity. With these new products, federal or state government may not be able to develop regulations rapidly enough to address each product and the nuances associated with it. Therefore, an approach is needed that establishes basic safety requirements, product identity, and removes government as the quality control department for a company or municipality. Industry has the primary responsibility for quality assurance and producing an unadulterated product. Government's role is one of oversight to ensure that industry is fulfilling its role.

To address the quality assurance issue, the FDA is suggesting that manufacturers implement a HACCP (Hazard Analysis Critical Control Points) program to address safety issues associated with their products. Good manufacturing practices (GMPs) have and continue to work effectively for specific areas in both human food and animal feed. Examples of successful application of GMPs are the food sanitary GMPs, outlined in 21 CFR, Part 110, and the GMPs' use in the medicated feed industry (21 CFR §225). In both instances, GMPs address specific problems and control points that are specific and common to all members of the industries to which they are applied. Because of the specific nature of GMPs, they are not particularly well suited to operations within an industry with great diversity or with many new products or product uses. Therefore, GMPs for the use of these nontraditional feed ingredients, while possible, are not practical because of the breadth and diversity of the industries and the resources within government that would be required to develop the GMPs.

In summary, we believe protection of public health is a goal that the FDA, the animal feed industry, and the animal producer all share. Our continued efforts in assuring compliance with the regulations are an example of the commitment that both government and industry have to that goal.

References

Animal Proteins Prohibited in Ruminant Feed. 1998. Title 21 Code of Federal Regulations §589.2000. U.S. Government Printing Office.

Association of American Feed Control Officials. 1999. Official Publication.

Current Good Manufacturing Practice for Medicated Feeds. 1999. Title 21 Code of Federal Regulations § 225. U.S. Government Printing Office.

Current Good Manufacturing Practice in Manufacturing, Packing, or Holding Human Food. 1998. Title 21 Code of Federal Regulations § 110. U.S. Government Printing Office.

Eligibility for Classification as Generally Recognized as Safe (GRAS). 1998. Title 21 Code of Federal Regulations § 570.30. U.S. Government Printing Office.

Federal Food, Drug, and Cosmetic Act as Amended. (FFDCA). 1998. § 201 (s). Department of Health and Human Services, Food and Drug Administration.

Federal Food, Drug, and Cosmetic Act as Amended. (FFDCA). 1998. § 201 (f). Department of Health and Human Services, Food and Drug Administration.

Federal Food, Drug, and Cosmetic Act as Amended. (FFDCA). 1998. Department of Health and Human Services, Food and Drug Administration.

Federal Meat Inspection Act (21 U.S.C. 601 et seq.).

50 Federal Register 27294-27296, July 2, 1985.

62 Federal Register 40570-40600, July 29, 1997.

62 Federal Register 552-566, January 3, 1997.

62 Federal Register 30936, June 5, 1997.

Food Additive Petitions. Title 21 Code of Federal Regulations § 571. U.S. Government Printing Office.

Food Additives. Title 21 Code of Federal Regulations § 570. U.S. Government Printing Office.

Poultry Products Inspection Act (21 U.S.C. 451 et seq.).

Public Health Service Act. (42 U.S.C. § 201 et seq.).

Substances That are Generally Recognized as Safe. Title 21 Code of Federal Regulations § 582.1. U.S. Government Printing Office.

Sulfiting Agents; Revocation of GRAS Status for Use on Fruits and Vegetables to be Served or Sold Raw to Consumers. 51 Federal Register 25021, July 9, 1986.

Swine Health Protection Act. 1998. Title 9 Code of Federal Regulations § 166. U.S. Government Printing Office.

Swine Health Protection, Definitions in Alphabetical Order. 1998. Title 9 Code of Federal Regulations §166.1. U.S. Government Printing Office.

U.S. verses An Article of Food *** Coco Rico, 752 F.2d 11 (1st Cir. 1985).

United States v. 50 Boxes * * * Cafergot P- B Suppositories, 721 F.Supp. 1462, 1465 (D. Mass. 1989), aff'd, 909 F.2d 24 (1st Cir. 1990); An Article of Drug * * * Furestrol Vaginal Suppositories, 251 F.Supp. 1307 (N.D. Ga. 1968), aff'd, 415 F.2d 390 (5th Cir. 1969), see also 5,906 Boxes, 745 F. 2d at 119 n.22.

Weinberger verses Hynson, Wescott and Dunning Inc., 412 U.S. 606, 632 (1073).

4

Regulation of Food-waste Feeding

by Roger D. Hoestenbach

Except for the Swine Health Protection Act previously discussed, the regulation of recycled food waste has very few differences from any other feedstuff. And, unless you are feeding swine, the Swine Health Protection Act does not apply.

The regulation of animal feeds in the United States is complicated and requires dealing with several different agencies, both state and federal. Unfortunately, there are no shortcuts. However, AAFCO is a source that can offer direction while saving time and expense.

"A basic goal of AAFCO is to provide a mechanism, for developing and implementing uniform and equitable laws, regulations, standards, definitions, and enforcement policies for regulating the manufacture, labeling, distribution, and sale of animal feeds, resulting in safe, effective, and useful feeds. The Association thereby promotes new ideas and innovative procedures and urges their adoption by member agencies, for uniformity (see page 65 AAFCO 1998 Official Publication)." This directive guides AAFCO in the general conduct of Association business. Among the business of AAFCO is providing model legislation, rules, and policies for the regulation of animal feed and definitions to use in describing that feed. AAFCO, while not a regulatory authority, represents the authority of its membership. The membership of AAFCO consists of officers charged with the execution of state, province, dominion, and federal laws regulating the production, labeling, distribution, and sale of animal feeds. While following the AAFCO model may not always ensure meeting the requirements for every state, it is the most universally acceptable approach to both state and federal requirements. Within the United States, contacting state control officials is the first and best source of information. They can provide the necessary assistance for distributing within their state, and they can also provide both guidance and contacts for complying with federal laws and marketing in other states.

In order to discuss feed regulation, feed should be defined first. The FFDCA (1998) defines food as, "articles used for food or drink for man or other animals, chewing gum, and articles used for components of any such article." Therefore, by definition, feed is food under the federal statutes.

AAFCO's model bill defines commercial feed as, "all materials or combination of materials which are distributed or intended for distribution for use as feed or for mixing in feed, unless such materials are specifically exempted." This usually will include vitamins, minerals, antibiotics, antioxidants, medicines, drugs, chemicals, organics, inorganics, or other materials used as ingredients or components of mixtures of materials used as feed for animals. Simply put, feed includes anything consumed orally, except for water, by animals other than man.

Labels are the cornerstones of any feed regulatory program. The identity of the product is its label. Labels are required to contain all the information necessary to distribute that feed and are the basis for most of the regulations. The label is in essence a contract between the distributor of the feed and its user. Therefore, the label should provide all the information necessary for the user to understand what the product is, what it is for, and how it should be used. In order to further define what an individual feed is, we can refer to Regulation 5(a) of the model regulations, which states, "The nutritional content of commercial feed shall be as purported or is represented to possess by its labeling. Such animal feed, its labeling, and intended use must be suitable for the intended purpose of the product."

These requirements establish what is generally required for proper labeling of most feeds and must be in the following format (figure 4.1):

(1) Brand name and product name
(2) Purpose statement
(3) Guaranteed analysis
(4) List of ingredients
(5) Directions for use and any warning or caution statements
(6) Name and address of manufacturer
(7) Quantity statement

Regulation 6(a) of the model regulations states: "The name of each ingredient or collective term for the grouping of ingredients, when required to be listed, shall be the name as defined in the Official Definitions of Feed Ingredients as published in the Official Publication of the Association Of American Feed Control Officials, the common or usual name, or one approved . . ."

Bluebird Beef Feed
FOR BEEF CATTLE ON PASTURE

Guaranteed Analysis
Crude protein (Min).............................12.0%
(This includes not more than 2.9% equivalent
crude protein from non-protein nitrogen)
Crude fat (Min)......................................2.0%
Crude fiber (Max)................................10.0%
Calcium (Min)...0.5%
Calcium (Max).......................................1.0%
Phosphorus (Min)..................................0.5%
Salt (Min)..11.0%
Salt (Max)..13.2%
Potassium (Min)......................................0.4%
Vitamin A (Min)......................10,000 IU/Lb

Ingredient Statement
Grain Products, Plant Protein Products, Cane Molasses, Dehydrated
Restaurant Food Waste, Processed Grain By-Products, Urea, Vitamin A
Supplement, Vitamin D3 Supplement, Vitamin E Supplement, Calcium
Carbonate, Monocalcium Phosphate, Salt, Manganous Oxide, Ferrous Sulfate,
Copper Oxide, Magnesium Oxide, Zinc Oxide, Cobalt Carbonate,
Ethylenediamine Dihydriodode, Potassium Chloride.

FEEDING DIRECTIONS
Self-feed to beef cattle on pasture. Feed 4–6 pounds per head per day as a pasture extender.

Provide plenty of fresh, clean water at all times.

Manufactured by
BlueBird Feed Mill
Anytown, Texas 77777
NET WT 50LB (22.6 Kg)

Figure 4.1. Animal feed label.

There are currently a number of feed terms and officially defined ingredients that have specific application in defining recycled food or processing wastes and need to be considered or would be included when regulating food waste. AAFCO terminology includes (see page 167-180 AAFCO 1998 Official Publication)

Refuse. (Part) Damaged, defective, or superfluous edible material produced during or leftover from a manufacturing or industrial process.

Sludge. The suspended or dissolved solid matter resulting from the processing of animal or plant tissue for human food. (Note: do not confuse with sewage sludge.)

Uncleaned. (Physical form) Containing foreign material.

Waste. (Part) See refuse.

(Current specific AAFCO definitions are within "60. Miscellaneous Products," *1998 Official Publication,* Investigator and Section Editor, Shannon Jordre, Program Specialist, South Dakota Department of Agriculture, pp. 250-255). Below are several food waste definitions.

60.15 Dried Bakery Product (IFN 4-00-466) is a mixture of bread, cookies, cake, crackers, flours, and dough that has been mechanically separated from nonedible material, artificially dried and ground. If a product contains more than 3.5% salt, the maximum percentage of salt must be part of the name; that is, Dried Bakery Product with _____ % Salt. (Proposed 1962, Adopted 1967.) (Bakery waste dehydrated.)

60.33 Dehydrated Food-Waste (IFN 4-12-175). Any and all animal and vegetable produce picked up from basic food processing sources or institutions where food is processed. The produce shall be picked up daily or sufficiently often so that no decomposition is evident. Any and all undesirable constituents shall be separated from the material. It shall be dehydrated to a moisture content of not more than 12% and be in a state free from all harmful microorganisms. (Proposed 1975, Adopted 1976.) (Food waste dehydrated).

60.12 Dehydrated Garbage (4-02-092) is composed of artificially dried animal and vegetable waste collected sufficiently often that harmful decomposition has not set in, and from which have been separated crockery, glass, metal, string, and similar materials. It must be processed at a temperature sufficient to destroy all organisms capable of producing animal diseases. If part of the grease and fat is removed, it must be designated as "Degreased Dehydrated Garbage."(Adopted 1954, Amended 1963.) (Garbage dehydrated.)

60.28 Dried Potato Products (IFN 4-03-775) is the dried residue of potato pieces, peelings, culls, etc., obtained from the manufacture of processed potato products for human consumption. The residue may contain up to 3% hydrate of lime, which may be added to aid in processing. (Proposed 1972, Adopted 1973.) (Potato process residue dehydrated.)

60.35 Sugar Foods By-product (IFN 4-20-865) is the product resulting from the grinding and mixing of the inedible portions derived from the preparation and packaging of sugar-based food products such as candy, dry packaged drinks, dried gelatin mixes, and similar food products that are largely sugar. It shall contain not less dm 80% total sugar expressed as invert. It shall be free from foreign materials harmful to animals. (Proposed 1976, Adopted 1977.) (Sugar foods process residue.)

The following are two new definitions currently being proposed to the AAFCO general membership for adoption.

T60.96 Food-Processing Waste is composed of any and all animal and vegetable products from basic food processing. This may include manufacturing or processing waste, cannery residue, production overrun, and otherwise unsaleable material. The guaranteed analysis shall include the maximum moisture, unless the product is dried by artificial means to less than 12% moisture and designated as "Dehydrated Food Processing Waste." If part of the grease and fat is removed, it must be designated as "Degreased."

T60.97 Restaurant Food Waste is composed of edible food waste collected from restaurants, cafeterias, and other institutes of food preparation. Processing and/or handling must remove any and all undesirable constituents including crockery, glass, metal, string, and similar materials. The guaranteed analysis shall include maximum moisture, unless the product is dried by artificial means to less than 12% moisture and designated as "Dehydrated Restaurant Food Waste." If part of the grease and fat is removed, it must be designated as "Degreased."

There are secondary concerns for efficacy, the availability of nutrients, and the presence of antinutrient properties. There are also valid public health concerns that include pathogens (both human and animal), toxins such as pesticides, industrial chemicals, or naturally occurring toxins such as mycotoxins from waste and improper storage of material, heavy metals, and nutritional or dietary imbalances. Standard to the adulteration language commonly contained in most state laws, such as stated in Section 7(a)(7) of the AAFCO model bill, "A commercial feed shall be deemed to be adulterated if it consists in whole or in part of any filthy, putrid, or decomposed substance, or if it is otherwise unfit for feed."

The need may arise for additional additives to balance or enhance the nutrition of the product. These might include the addition of minerals for balance or to offset a deficiency or antinutritive effect, the addition of sequestering or chelating agents, or the use of a variety of preservatives. There are also potential problems when remediating wastes with the utilization of biosolids in that the levels of compounds present may favor a particular microbe that is undesirable, or at least less desirable, or there may be very hostile growth conditions due to pH or other physical conditions. Even the most benign of microbes potentially may give rise to a pathogenic strain whenever there are large populations present. The storage processes must be carefully monitored whenever there are live organisms involved, and steps must be taken to ensure the stability of the material. It can deteriorate just as anything else used in animal feeds.

Food processors generate far more organic wastes than many others do because they deal exclusively with consumables. Spurred by the Clean Water Act and other similar environmentally friendly legislation, mandates were created to reduce the amount of organics in both land-fillable items and water for sewage treatment. Because of the increased costs of disposal and, in some cases, the increasing difficulties in finding sites to accept organic products, many producers are looking for alternative uses. These may include recycling back into processing, such as what occurs with water removed during processing from organics by dewatering. This not only reduces the need for additional water in manufacturing, but also allows for reduction of the volume and weight of any disposable material. Usually the deciding factor when considering options is the location of market. Transportation costs will determine what form and how to best address recycling needs. There are economic incentives when landfilling is averaging in excess of $100/ton east of the Mississippi and even low-profit or no-profit items can result in positive cash flow when compared with traditional disposal. While it would be best to show an actual profit, simply increasing efficiency and reducing the high cost of disposal may be the key to containing the wastes. The problems of recycling are addressable within a quality control program and within the proper design of waste handling systems. However, they may require combining techniques of composting for developing fertilizer and/or feed and isolation of the components using technologies from a variety of areas in order to become economically practical.

The technologies also exist that would allow the use of biosolids produced in the remediation of even common septic/sanitary sewage waste as animal feeds. It would not be without problems including pathogens such as *E. coli* and *Salmonella,* magnification of metals, not just so called "heavy metals," but nutritional sources that may become toxic or at least antinutritive by competitive interference at levels currently encountered in sewage sludge. These septic/sanitary sewage wastes are mentioned because they are frequently encountered in the grease rendering industry improperly collected from restaurants. These improper collections are usually from the sewer grease traps required by municipalities for restaurants and other food handling facilities. Rendered products containing material from sanitary sewage grease traps are not currently allowed to be used in animal feed. However, there are acceptable grease traps for the collections of waste greases for use in animal feed and they are frequently found in basic food manufacturing facilities. The key difference in design that allows their acceptablility is isolation from contaminate sanitary sewage sources. These precautions are not usually present in conventional restaurant systems.

The regulation of recycled food waste is complicated and often confusing. However, there are numerous agencies and contacts that can help in answering questions and assisting in the proper use and introduction of these products into the marketplace. (See Appendix A)

References

Federal Food, Drug, and Cosmetic Act, As Amended. (FFDCA) 1998. United Sates Code, Title 21. U.S. Government Printing Office, Washington, D.C. Official Publication of The Association of American Feed Control Officials (1998); Paul Bachman, Editor; St. Paul, Minnesota.

Personal correspondence and general discussion notes with the AAFCO Environmental Issues Committee.

Stack, Charles R. and Prasad S. Kodukula (April 24,1995); Production of Food Processing Biosolids and Their Use as Animal Feed: An Overview; Presentation to the AAFCO Environmental Issues Committee; Indianapolis, Indiana.

Wagner, Matina. 1995. Resource Recycling. Solid Waste Association of North America (SWANA). Silver Spring, Maryland.

5

The History and Enforcement of the Swine Health Protection Act

by Arnold C. Taft, Ernest W. Zirkle,
and Bonnie A. Altizio

The Swine Health Protection Act was developed with the intention of protecting the $7 billion dollar-a-year swine industry by regulating the feeding of garbage to pigs. It is enforced by the USDA, Animal Plant Health Inspection Service (APHIS), and Veterinary Services (VS). The bill was sponsored by Representatives Paul Findley and Edward Madigan, both from Illinois, on February 25, 1980, and signed into law on October 17, 1980 (Congressional Record 1980; Congressional Quarterly Weekly Report 1980). The Act is aimed at keeping certain foreign diseases such as foot-and-mouth disease (FMD), African swine fever (ASF), vesicular exanthema of swine (VES), and hog cholera (HC) out of the United States. These diseases may be transmitted by feeding raw or partially-cooked infected tissues to swine. An official of the USDA testified that it is an established fact that infections and communicable swine diseases can be spread readily and rapidly if swine are fed garbage that is either raw or improperly treated to kill disease organisms (House Reports 1980). The virus causing ASF survives in undercooked pork from infected hogs for long periods of time and can remain infectious unless heated at very high temperatures (House Reports 1980). Thus, garbage, as a source of food, is a vital link in the transmission of disease to swine.

Although FMD is found in other major livestock-producing countries of the world, it is not found in North America today. The United States experienced nine outbreaks between 1870 and 1929 (Mohler 1929; 1938 Mohler and Traum 1942). In all cases but two, the disease was eradicated within a few months. The two outbreaks in 1914 to 1916 and 1924 to 1925 in California took 20 months of concentrated effort until the areas were clean and restrictions were lifted. The 1914 to 16 outbreak resulted in the slaughter of 170,000 swine, cattle, sheep, and goats. During the

California outbreak in 1924 to 1925, more than 130,000 swine, cattle, deer, sheep, and goats were destroyed due to infection or exposure to the disease. FMD was eradicated on March 18, 1929 (Callis et al. 1975).

Canada and Mexico also have been affected by FMD. In Canada, it was diagnosed in 1952, and the USDA considered the country infection-free a year later (Childs 1952). According to Law (1915), it spread to Montreal and then into New York via Ontario and Quebec in the late 1800s. FMD was confirmed in Mexico in 1946 (USDA Release 1947), and it was not until six years later that the country was declared disease-free (Shahan 1954). FMD appeared again in the spring of 1953 (USDA Release 1953), and the USDA restricted the importation of cattle, swine, sheep, and goats from Mexico into the United States until 1954 (USDA Release, 1954a,b). It has not been detected since the mid-1950s (Mexico–United States Commission for the Prevention of Foot-and-Mouth Disease 1972).

African swine fever has been present for many years in eastern and southern Africa. According to Ribeiro et al. (1958) and Ribeiro and Azevedo (1961), ASF spread to Portugal in 1957, where nearly 17,000 swine were exposed. During 1957 and 1958, nearly 6,500 animals died, and the remaining animals were slaughtered to eradicate the disease. In 1960, ASF was found in Portugal again, and another 14,000 animals were killed. The disease also reached Spain. Approximately 30% of the outbreaks were a result of feeding uncooked garbage to swine. By the spring of 1961, this disease had spread to France, but was eliminated by slaughtering all swine, regardless of confirmed infection. In 1967, ASF was located in Italy, a strict slaughter policy was instituted, and the cooking of garbage was made compulsory. A small outbreak occurred in 1968 that was controlled. ASF was first diagnosed in the Western Hemisphere in 1971. Although ASF has never reached the United States, its presence in Africa and Europe and its ability to spread easily makes it an ongoing threat to the world swine population (Mauer 1975).

In 1932, a swine herd at Buena Park, California (50 miles east of Los Angeles), became infected with what was thought to be FMD. The government required quarantine and inspection. It was later confirmed to be VES, a new disease (Madin 1975). Within the next week, it spread to five other ranches within a 15-mile radius. About two weeks after the detection of the first outbreak, it was found 50 miles away in San Bernardino County. This was the last reported outbreak in 1932. All animals directly or indirectly related were slaughtered. The state of California and the federal government paid more than $200,000 for the loss of about 18,025 animals. All three herds had the common factor of being fed raw garbage. The food was obtained from restaurants that served a variety of fish products. From 1935 to 1944, VES appeared in California annually, involving 430,000 animals, which totaled more than 40% of that state's swine population. The disease agent was also identified in various marine mammals off the California coast. This disease was confined to California until 1952 when it spread to Wyoming (Madin 1975), as a result of feeding scraps of infected pork that originated from an interstate California passenger train to pigs. Before the detection of VES, some pigs from Wyoming were shipped to Nebraska. It was an epidemic that lasted for five years and involved 42 states and the District of Columbia before its eradication in 1959. It cost the federal government $33 million, including indemnity charges (Mulhern 1953). The final cases of VES were found on three large swine operations located in Secaucus, New Jersey, where garbage was being fed to pigs. The eradication effort and some state laws requiring the cooking of garbage made the control of VES possible. In 1959, VES was declared an exotic disease in the United States by the Secretary of Agriculture (House and House 1992).

In 1962, a cooperative state and federal HC eradication program began. The last outbreak of HC in the United States occurred in 1976. The total cost to eradicate HC from the United States amounted to about $140 million (Van Oirschot 1992). An important source of HC infection was discovered to be virus-containing garbage that had not been properly sterilized. This mode of virus spread was responsible for 18% of the cases of HC in 1972 and 22% of those during 1973 (Dunne 1975). The United States was declared free of HC in 1978.

As a part of the hog cholera eradication program, the USDA placed restrictions on interstate movement of raw, garbage-fed swine and their pork products. These restrictions state that swine fed any raw garbage may be shipped interstate for immediate slaughter and special processing, and products derived from raw garbage fed swine must be specially processed prior to interstate shipment (House Reports 1980).

Regulation

Improperly cooked garbage that is fed to swine has been linked to the spread of several diseases. To reduce this risk and protect the swine industry, Congress passed the Swine Health Protection Act to regulate the feeding of garbage to swine. (The full text of the Act, amended as of 1998, is included in Appendix B.) Included below is the current text of the Code of Federal Regulations, Title 9, Sections 166 and 167, revised as of January 1, 1998. The code is the regulatory interpretation of the Act and sets the law for the feeding of garbage to swine, such as standards for cooking, licensing, recordkeeping, etc. Also defined are the status of individual states and whether the federal government or the individual state has primary enforcement responsibilities.

Code of Federal Regulations

TITLE 9—ANIMALS AND ANIMAL PRODUCTS

CHAPTER I—ANIMAL AND PLANT HEALTH INSPECTION SERVICE, DEPARTMENT OF AGRICULTURE

SUBCHAPTER L—SWINE HEALTH PROTECTION

PART 166—SWINE HEALTH PROTECTION

General Provisions

Sec.
166.1   Definitions in alphabetical order.
166.2   General restrictions.
166.3   Separation of swine from the garbage handling and treatment areas.
166.4   Storage of garbage.
166.5   Licensed garbage-treatment facility standards.
166.6   Swine feeding area standards.
166.7   Cooking standards.
166.8   Vehicles used to transport garbage.
166.9   Recordkeeping.
166.10 Licensing.
166.11 Suspension and revocation of licenses.
166.12 Cancellation of licenses.
166.13 Licensee responsibilities.
166.14 Cleaning and disinfecting.
166.15 State status.

Authority: 7 U.S.C. 3802, 3803, 3804, 3808, 3809, and 3811; 7 CFR 2.22, 2.80, and 371.2(d).

Source: 47 FR 49945, Nov. 3, 1982, unless otherwise noted.

General Provisions

Sec. 166.1 Definitions in alphabetical order.

For the purposes of this part, the following terms shall have the meanings assigned them in this section. Unless otherwise required by the context, the singular form shall also import the plural and the masculine form shall also import the feminine, and vice versa. Words undefined in the following paragraphs shall have the meaning attributed to them in general usage as reflected by definitions in a standard dictionary.

Act. The Swine Health Protection Act (Pub. L. 96-468) as amended by the Farm Credit Act Amendments of 1980 (Pub. L. 96-592).

Administrator. The Administrator, Animal and Plant Health Inspection Service, or any person authorized to act for the Administrator.

Animal and Plant Health Inspection Service (APHIS). The Animal and Plant Health Inspection Service of the United States Department of Agriculture.

Animals. All domesticated and wild mammalian, poultry, and fish species, and wild and domesticated animals, including pets such as dogs and cats.

Area Veterinarian in Charge. The veterinarian of APHIS who is assigned by the Administrator to supervise and perform the official work of APHIS in a State or States or any other official to whom authority has heretofore been delegated or to whom authority may hereafter be delegated to act in his stead.

Facility. The site and all objects at this site including equipment and structures where garbage is accumulated, stored, handled, and cooked as a food for swine and which are fenced in or otherwise constructed so that swine are unable to have access to untreated garbage.

Garbage. All waste material derived in whole or in part from the meat of any animal (including fish and poultry) or other animal material, and other refuse of any character whatsoever that has been associated with any such material, resulting from the handling, preparation, cooking or consumption of food, except that such term shall not include waste from ordinary household operations which is fed directly to swine on the same premises where such household is located.

Inspector. Any individual employed by the United States Department of Agriculture or by a State for the purposes of enforcing the Act and this part.

License. A permit issued to a person for the purpose of allowing such person to operate a facility to treat garbage that is to be fed to swine.

Licensee. Any person licensed pursuant to the Act and regulations.

Person. Any individual, corporation, company, association, firm, partnership, society or joint stock company or other legal entity.

Premises. The location of a garbage treatment facility, as defined in this part, and any areas owned or controlled by the operator of the facility where swine are kept or fed by the operator.

Rendered product. Waste material derived in whole or in part from the meat of any animal (including fish and poultry) or other animal material, and other refuse of any character whatsoever that has been associated with any such material, resulting from the handling, preparation, cooking or consumption of food that has been ground and heated to a minimum temperature of 230 deg. F. to make products such as, but not limited to, animal, poultry, or fish protein meal, grease or tallow.

State. The fifty States, the District of Columbia, Guam, Puerto Rico, the Virgin Islands of the United States, American Samoa, the Commonwealth of the Northern Mariana Islands, and the territories and possessions of the United States.

State animal health official. The individual employed by a State who is responsible for livestock and poultry disease control and eradication programs or any other official to whom authority is delegated to act for the State animal health official.

Treated garbage. Edible waste for animal consumption derived from garbage (as defined in this section) that has been heated throughout at boiling or equivalent temperature (212 deg. F. or 100 deg. C. at sea level) for 30 (thirty) minutes under the supervision of a licensee.

Treatment. The heating of garbage to specifications as set forth in this part.

Untreated garbage. Garbage that has not been treated in accordance with the Act and these regulations.

(Sec. 511, Pub. L. 96-592, 94 Stat. 3451 (7 U.S.C. 3802); secs. 4, 5, 9,12, Pub. L. 96-468, 94 Stat. 2229 (7 U.S.C. 3803, 3804, 3808, 3811) 7CFR 2.17, 2.51, and 371.2(d))

[47 FR 49945, Nov. 3, 1982, as amended at 48 FR 22290, May 18, 1983; 52FR 4890, Feb. 18, 1987; 56 FR 26899, June 12, 1991]

Sec. 166.2 General restrictions.

(a) No person shall feed or permit the feeding of garbage to swine unless the garbage is treated to kill disease organisms, pursuant to this Part, at a facility operated by a person holding a valid license for the treatment of garbage; except that the treatment and license requirements shall not apply to the feeding or the permitting of the feeding to swine of garbage only because the garbage consists of any of the following: rendered products; bakery waste; candy waste; eggs; domestic dairy products (including milk); fish from the Atlantic Ocean within 200 miles of the continental United States or Canada; or fish from inland waters of the United States or Canada which do not flow into the Pacific Ocean.

(b) No person operating such a facility may be licensed to treat garbage unless he or she meets the requirements of this part designed to prevent the introduction or dissemination of any infectious or communicable disease of animals and unless the facility is so constructed that swine are unable to have access to untreated garbage or equipment and material coming in contact with untreated garbage.

(c) The regulations of this part shall not be construed to repeal or supersede State laws that prohibit feeding of garbage to swine or to prohibit any State from enforcing requirements relating to the treatment of garbage that is to be fed to swine or the feeding thereof which are more stringent than the requirements contained in this part. In a State which prohibits the feeding of garbage to swine, a license under the Act will not be issued to any applicant.

(Sec. 511, Pub. L. 96-592, 94 Stat. 3451 (7 U.S.C. 3802); secs. 4, 5, 9,10, 12, Pub. L. 96-468, 94 Stat. 2229-2233 (7 U.S.C. 3803, 3804, 3808,3809, 3811) 7 CFR 2.17, 2.51, and 371.2(d))

[47 FR 49945, Nov. 3, 1982, as amended at 49 FR 14497, Apr. 12, 1984; 52FR 4890, Feb. 18, 1987]

Sec. 166.3 Separation of swine from the garbage handling and treatment areas.

(a) Access by swine to garbage handling and treatment areas shall be prevented by construction of facilities to exclude all ages and sizes of swine.
(b) All areas and drainage therefrom, used for the handling and treatment of untreated garbage shall be inaccessible to swine on the premises. This shall include the roads and areas used to transport and handle untreated garbage on the premises.

Sec. 166.4 Storage of garbage.

(a) Untreated garbage at a treatment facility shall be stored in covered and leakproof containers until treated.
(b) Treated garbage shall be transported to a feeding area from the treatment facility only in
   (1) Containers used only for such treated garbage;
   (2) Containers previously used for garbage which have been cleaned and disinfected in accordance with Sec. 166.14 of this part; or
   (3) Containers in which the garbage was treated.

[47 FR 49945, Nov. 3, 1982, as amended at 52 FR 4890, Feb. 18, 1987]

Sec. 166.5 Licensed garbage-treatment facility standards.

Garbage-treatment facilities shall be maintained as set forth in this section.

(a) Insects and animals shall be controlled. Accumulation of any material at the facility where insects and rodents may breed is prohibited.
(b) Equipment used for handling untreated garbage, except for the containers in which the garbage has been treated, may not be subsequently used in the feeding of swine unless first cleaned and disinfected as set forth in Sec. 166.14(b).
(c) Untreated garbage that is not to be fed to swine and materials in association with such garbage shall be disposed of in a manner

consistent with all applicable governmental environmental regulations and in an area inaccessible to swine.

[47 FR 49945, Nov. 3, 1982, as amended at 52 FR 4890, Feb. 18, 1987]

Sec. 166.6 Swine feeding area standards.

Untreated garbage shall not be allowed into swine feeding areas. Any equipment or material associated with untreated garbage, except for containers holding treated garbage which was treated in such containers, shall not be allowed into swine feeding areas at treatment premises until properly cleaned and disinfected as set forth in Sec. 166.14(b) of this part.

[47 FR 49945, Nov. 3, 1982, as amended at 52 FR 4890, Feb. 18, 1987]

Sec. 166.7 Cooking standards.

(a) Garbage shall be heated throughout at boiling (212 deg. F. or 100 deg. C. at sea level) for 30 (thirty) minutes.
(b) Garbage shall be agitated during cooking, except in steam cooking equipment, to ensure that the prescribed cooking temperature is maintained throughout the cooking container for the prescribed length of time.

Sec. 166.8 Vehicles used to transport garbage.

Vehicles used by a licensee to transport untreated garbage, except those that have also been used to treat the garbage so moved, shall not be used for hauling animals or treated garbage until cleaned and disinfected as set forth in Sec. 166.14(c) of this part.

[47 FR 49945, Nov. 3, 1982, as amended at 52 FR 4890, Feb. 18, 1987]

Sec. 166.9 Recordkeeping.

(a) Each licensee shall record the destination and date of removal of all treated or untreated garbage removed from the licensee's premises.
(b) Such records shall be legible and indelible.
(c) Each entry in a record shall be certified as correct by initials or signature of the licensee or an authorized agent or employee of the licensee.
(d) Such records shall be maintained by the licensee for a period of 1 year from the date made and shall be made available to inspectors upon request during normal business hours at that treatment facility.

(Approved by the Office of Management and Budget under control number 0579-0066)

[47 FR 49945, Nov. 3, 1982, as amended at 48 FR 57474, Dec. 30, 1983; 52FR 4890, Feb. 18, 1987]

Sec. 166.10 Licensing.

(a) Application. Any person operating or desiring to operate a treatment facility for garbage that is to be treated and fed to swine shall apply for a license on a form which will be furnished, upon request, by the Area Veterinarian in Charge or, in States with primary enforcement responsibility, by the State animal health official in the State in which the person operates or intends to operate. When a person operates more than one treatment facility, a separate application to be licensed shall be made for each facility. Exemptions to the requirements of this paragraph may be granted in States other than those with primary enforcement responsibility by the Administrator, if he finds that there would not be a risk to the swine industry in the United States. Any person operating or desiring to operate a facility to treat garbage to be fed to swine who would otherwise be required under this part to obtain a license to treat garbage only because it contains one or more of the items allowed to be fed to swine under Sec. 166.2(a) of this part is exempted from the requirements of this paragraph.

(b) Acknowledgement of Act and regulations. A copy of the Act and regulations shall be supplied to the applicant at the time the applicant is given a license application. The applicant shall sign a receipt at the time of the prelicensing inspection acknowledging that the applicant has received a copy of the Act and regulations, that the applicant understands them, and agrees to comply with the Act and regulations.

(c) Demonstration of compliance with the regulations.

  (1) Prior to licensing, each applicant shall demonstrate during an inspection of the premises, facilities, and equipment that the facilities and equipment to be used in the treatment of garbage comply with these regulations. If the applicant's facilities and equipment do not meet the standards established by the regulations, the applicant shall not be licensed and shall be advised of the deficiencies and the measures that must be taken to comply with the regulations.

  (2) The licensee shall make the premises, facilities, and equipment available during normal business hours for inspections by an inspector to determine continuing compliance with the Act and regulations.

  (3) The facilities and equipment of an applicant for a license shall be in compliance with all applicable governmental environmental regulations before the applicant will be licensed.

(d) Issuance of license. A license will be issued to an applicant when the requirements of paragraphs (a), (b), and (c) of this section have been met, provided that such facility is not located in a State which prohibits the feeding of garbage to swine; and further, that if the Administrator has reason to believe that the applicant for a license is unfit to engage in the activity for which application has been made by reason of the fact that the applicant is engaging in or has, in the past, engaged in any activity in apparent violation of the Act or the regulations which has not been the subject of an administrative proceeding under the Act, an administrative proceeding shall be promptly instituted in which the applicant will be afforded an opportunity for a hearing in accordance with the rules of practice under the Act, for the purpose of giving the applicant an opportunity to show cause why the application for license should not be denied. In the event it is determined that the application should be denied, the applicant shall be precluded from reapplying for a license for 1 year from the date of the order denying the application.

(Approved by the Office of Management and Budget under control number 0579-0065)

(Sec. 511, Pub. L. 96-592, 94 Stat. 3451 (7 U.S.C. 3802); secs. 4, 5, 9,10, 12, Pub. L. 96-468, 94 Stat. 2229-2233 (7 U.S.C. 3803, 3804, 3808,3809, 3811) 7 CFR 2.17, 2.51, and 371.2(d) [47 FR 49945, Nov. 3, 1982, as amended at 48 FR 57474, Dec. 30, 1983; 49 FR 14497, Apr. 12, 1984; 52 FR 4890, Feb. 18, 1987; 56 FR 26899, June 12, 1991]

Sec. 166.11 Suspension and revocation of licenses.

(a) Suspension or revocation after notice. In addition to the imposition of civil penalties and the issuance of cease and desist orders under the Act, the license of any facility may be suspended or revoked for any violation of the Act or the regulations in this part. Before such action is taken, the licensee of the facility will be informed in writing of the reasons for the proposed action and, upon request, shall be afforded an opportunity for a hearing with respect to the merits or validity of such action, in accordance with rules of practice which shall be adopted for the proceeding.

(b) Summary suspension. If the Administrator has reason to believe that any licensee has not complied or is not complying with any provisions of the Act or regulations in this part and the Administrator deems such action necessary in order to protect the public health, interest, or safety, the Administrator may summarily suspend the license of such persons pending a final determination in formal proceedings and any judicial review thereof, effective upon verbal or written notice of such suspension and the reasons therefor. In the event of verbal notification, written confirmation shall follow as soon as circumstances permit. This summary suspension shall continue in effect pending the completion of the proceeding and any judicial review thereof, unless otherwise ordered by the Administrator.

(c) The license of a person shall be automatically revoked, without action of the Administrator, upon the final effective date of the second criminal conviction of such person, as is stated in section 5(c) of the Act.

The licensee will be notified in writing of such revocation by the Area Veterinarian in Charge or, in States having primary enforcement responsibility, by the State animal health official.

(d) Any person whose license has been suspended or revoked for any reason shall not be licensed in such person's own name or in any other manner, nor shall any of such person's employees be licensed for the purpose of operating the facility owned or operated by said licensee while the order of suspension or revocation is in effect. Any person whose license has been revoked shall not be eligible to apply for a new license for a period of 1 year from the effective date of such revocation. Any person who desires the reinstatement of a license that has been revoked must follow the procedure for new licensees set forth in Sec. 166.10 of this part. [47 FR 49945, Nov. 3, 1982, as amended at 52 FR 4890, Feb. 18, 1987; 56 FR 26899, June 12, 1991]

Sec. 166.12 Cancellation of licenses.

(a) The Area Veterinarian in Charge or, in States listed in Sec. 166.15(d) of this part, the State animal health official shall cancel the license of a licensee when the Area Veterinarian in Charge or, in States listed in Sec. 166.15(d) of this part, the State animal health official finds that no garbage has been treated for a period of 4 consecutive months at the facility operated by the licensee. Before such action is taken, the licensee of the facility will be informed in writing of the reasons for the proposed action and be given an opportunity to respond in writing. In those instances where there is a conflict as to the facts, the licensee shall, upon request, be afforded a hearing in accordance with rules of practice which shall be adopted for the proceeding.

(b) Any licensee may voluntarily have his or her license canceled by requesting such cancellation in writing and sending such request to the Area Veterinarian in Charge, \1\ or, in States listed in Sec. 166.15(d) of this part, to the State animal health official. The Area Veterinarian in Charge or, in States listed in Sec. 166.15(d) of this part, the State animal health official shall cancel such license and shall notify the licensee of the cancellation in writing.

\1\ The name and address of the Area Veterinarian in Charge may be obtained from the Veterinary Services, Operational Support, 4700 River Road, Unit 33, Riverdale, Maryland 20737-1231.

(c) Any person whose license is canceled in accordance with paragraph (a) or (b) of this section may apply for a new license at any time by following the procedure for obtaining a license set forth in Sec. 166.10 of this part.

[52 FR 4891, Feb. 18, 1987, as amended at 56 FR 26899, June 12, 1991; 59FR 67618, Dec. 30, 1994]

Sec. 166.13 Licensee responsibilities.

(a) A licensed facility shall be subject to inspections. Each inspector will be furnished with an official badge or identification card, either of which shall be sufficient identification to entitle access during normal business hours to the facility for the purposes of inspection. At such time the inspector is duly authorized to:
    (1) Inspect the facility, including cooker function;
    (2) Take samples of garbage;
    (3) Observe and physically inspect the health status of all species of animals on the premises;
    (4) Review records and make copies of such records; and
    (5) Take photographs. A copy of each photograph will be provided to the licensee within 14 days.
(b) A licensee shall notify an inspector immediately upon detection of illness or death not normally associated with the licensee's operation in any animal species on the licensee's premises.
(c) A licensee shall notify an inspector or the State animal health official or the Area Veterinarian in Charge, as appropriate, of any change in the name, address, management or substantial control or ownership of his business or operation within 30 days after making such change.
(d) A licensee shall supply, upon request by an inspector, information concerning sources of garbage. Such information shall include the dates of supply and the names and addresses of the person and/or organization from which the garbage was received.

(Approved by the Office of Management and Budget under control number 0579-0065)

[47 FR 49945, Nov. 3, 1982, as amended at 48 FR 57474, Dec. 30, 1983; 52FR 4890, Feb. 18, 1987. Redesignated at 52 FR 4891, Feb. 18, 1987]

Sec. 166.14 Cleaning and disinfecting.

(a) Disinfectants to be used. Disinfection required under the regulations in this Part shall be performed with one of the following:
    (1) A permitted brand of sodium orthophenylphenate that is used in accordance with directions on the Environmental Protection Agency (EPA) approved label.
    (2) A permitted cresylic disinfectant that is used in accordance with directions on the EPA-approved label, provided such disinfectant also meets the requirements set forth in Secs. 71.10(b) and 71.11 of this chapter.
    (3) Disinfectants which are registered under the Federal Insecticide, Fungicide, and Rodenticide Act (7 U.S.C. 135 et seq.), with tuberculocidal claims and labeled as efficacious against any species within the viral genus Herpes, that are used for purposes of this Part in accordance with directions on the EPA-approved label.

(b) All premises at which garbage has been fed to swine in violation of the Act or regulations in this part shall, prior to continued use for swine feeding purposes, be cleaned and disinfected under the supervision of an inspector or an accredited veterinarian as defined in Part 160 of this chapter as follows: Empty all troughs and other feeding and watering appliances, remove all litter, garbage, manure, and other organic material from the floors, posts, or other parts of such equipment, and handle such litter, garbage, manure, and other organic material in such manner as not to allow animal contact with such material; clean all surfaces with water and detergent and saturate the entire surface of the equipment, fencing, troughs, chutes, floors, walls, and all other parts of the facilities, with a disinfectant prescribed in paragraph (a) of this section. An exemption to the requirements of this paragraph may be given by the Administrator or, in States with primary enforcement responsibility, by the State animal health official, when it is determined that a threat to the swine industry does not exist.

(c) Any vehicle or other means of conveyance and its associated equipment which has been used by the licensee to move garbage, except any vehicle or other means of conveyance which also has been used to treat the garbage so moved, shall, prior to use for livestock-related or treated garbage hauling purposes, be cleaned and disinfected as follows: Remove all litter, garbage, manure, and other organic material from all portions of each means of conveyance, including all ledges and framework inside and outside, and handle such litter, garbage, manure, and other organic material in such manner as not to allow animal contact with such material; clean the interior and the exterior of such vehicle or other means of conveyance and its associated equipment with water and detergent; and saturate the entire interior surface, including all doors, endgates, portable chutes, and similar equipment with a disinfectant prescribed in paragraph (a) of this section.

(d) The owner of such facilities and vehicles shall be responsible for cleaning and disinfecting as required under this section, and the cleaning and disinfecting shall be done without expense to the United States Department of Agriculture.

[47 FR 49945, Nov. 3, 1982. Redesignated and amended at 52 FR 4891, Feb.18, 1987; 56 FR 26899, June 12, 1991]

Sec. 166.15 State status.

(a) The following States prohibit the feeding of garbage to swine: Alabama, Delaware, Georgia, Idaho, Illinois, Indiana, Iowa, Louisiana, Mississippi, Nebraska, New York, North Dakota, South Carolina, South Dakota, Tennessee, Virginia, and Wisconsin.

(b) The following States and Puerto Rico permit the feeding of treated garbage to swine: Alaska, Arizona, Arkansas, California, Colorado,

Connecticut, Florida, Hawaii, Kansas, Kentucky, Maine, Maryland, Massachusetts, Michigan, Minnesota, Missouri, Montana, Nevada, New Hampshire, New Jersey, New Mexico, North Carolina, Ohio, Oklahoma, Oregon, Pennsylvania, Puerto Rico, Rhode Island, Texas, Utah, Vermont, Washington, West Virginia, and Wyoming.

(c) The following States have primary enforcement responsibility under the Act: Alabama, Arizona, California, Colorado, Florida, Georgia, Hawaii, Idaho, Illinois, Indiana, Iowa, Kansas, Michigan, Minnesota, Mississippi, Missouri, Montana, Nebraska, Nevada, New Jersey, New York, North Dakota, Ohio, Oregon, Pennsylvania, South Carolina, South Dakota, Tennessee, Utah, Virginia, West Virginia, and Wisconsin.

(d) The following States issue licenses under cooperative agreements with the Animal and Plant Health Inspection Service, but do not have primary enforcement responsibility under the Act: Kentucky, Maryland, Puerto Rico, Texas, and Washington.

(e) The public may contact the Area Veterinarian in Charge, Animal and Plant Health Inspection Service, United States Department of Agiculture or State animal health official, or the Animal and Plant Health Inspection Service, Veterinary Services, Swine Health, 4700 River Road, Unit 37, Riverdale, Maryland 20737-1231, concerning the feeding of garbage to swine.

[47 FR 49945, Nov. 3, 1982, as amended at 51 FR 2348, Jan. 16, 1986; 51FR 15757, Apr. 28, 1986. Redesignated and amended at 52 FR 4891, Feb.18, 1987. 52 FR 13231, Apr. 22, 1987; 52 FR 34208, Sept. 10, 1987; 52 FR37283, Oct. 6, 1987; 55 FR 30688, July 27, 1990; 56 FR 7555, Feb. 25,1991; 56 FR 26899, June 12, 1991; 56 FR 37827, Aug. 9, 1991; 59 FR67618, Dec. 30, 1994]

PART 167—RULES OF PRACTICE GOVERNING PROCEEDINGS UNDER THE SWINE HEALTH PROTECTION ACT

Subpart A—General

Sec.
167.1 Scope and applicability of rules of practice.

Subpart B—Supplemental Rules of Practice

167.10 Stipulations.

Authority: Sec. 5, 94 Stat. 2230; sec. 6, 94 Stat. 2231; sec. 12, 94 Stat. 2233; 7 U.S.C. 3804, 3805, 3811; 7 CFR 2.22, 2.80, 371.2(d).

Source: 48 FR 30095, June 30, 1983, unless otherwise noted.

Subpart A—General

Sec. 167.1 Scope and applicability of rules of practice.

The Uniform Rules of Practice for the Department of Agriculture promulgated in subpart H of part 1, subtitle A, title 7, Code of Federal Regulations, are the Rules of Practice applicable to adjudicatory, administrative proceedings under sections 5 and 6 of the Swine Health Protection Act (7 U.S.C. 3804, 3805). In addition, the Supplemental Rules of Practice set forth in subpart B of this part shall be applicable to such proceedings.

Subpart B—Supplemental Rules of Practice

Sec. 167.10 Stipulations.

(a) At any time prior to the issuance of a complaint seeking a civil penalty under the Act, the Administrator, in his discretion, may enter into a stipulation with any person in which:
  (1) The Administrator or the Administrator's delegate gives notice of an apparent violation of the Act, or the regulations issued there under, by such person and affords such person an opportunity for a hearing regarding the matter as provided by the Act;
  (2) Such person expressly waives hearing and agrees to a specified order which may include an agreement to pay a specified penalty within a designated time; and
  (3) The Administrator agrees to accept the order in settlement of the particular matter conditioned upon timely payment of the penalty if the order includes an agreement to pay a penalty.
(b) If the order includes an agreement to pay a penalty and the penalty is not paid within the time designated in such a stipulation, the amount of the penalty shall not be relevant in any respect to the penalty which may be assessed after issuance of a complaint.

Conclusion

In title 9, Code of Federal Regulations, part 166 section 166.15—State status—33 states plus Puerto Rico are identified as allowing garbage feeding on licensed premises. The remaining 17 states are listed that prohibit the feeding of garbage. Primary enforcement responsibilities under the Act are maintained by 32 states. The remaining states, plus Puerto Rico, have primary enforcement by the federal government.

All states make a report of swine health protection program activities four times a year. For the fourth quarter of 1997 (Oct, Nov., Dec. '97), 4,057 premises were licensed to feed garbage. During this reporting period, 3,820 of these premises were inspected. Through a variety of methods, 7,413 searches were made for noncompliant garbage feeders. Many of these searches were completed by other agencies that might identify a restaurant or food-service business that disposes of garbage to an unlicensed garbage feeder. The discovery of 392 violations, and all but four corrected, were reported during the period.

The pig is a very efficient animal through which to recycle food waste (garbage); however, such feed must be sterilized properly to prevent the introduction of a foreign disease. The Swine Health Protection Act has given us some minimum standards that if followed properly, will prevent introduction of a foreign disease. In order to ensure that these standards are met, perhaps it is time to raise the standards for cooking equipment, monitoring equipment for verification, and management of facilities licensed to feed food waste directly to swine.

New technologies are becoming available that can recycle food waste in large quantities, enabling them to have a shelf life and be properly sterilized. Food waste recycling symposiums have been sponsored by the New Jersey Department of Agriculture, Rutgers University Cooperation Extension, and the USDA in 1996-1999. These symposiums have initiated an emerging industry to recycle food waste.

Questions still remain to be answered, for example, (1) Who will regulate this emerging industry; (2) What are appropriate times, temperatures, and pressures to ensure a safe product; and (3) A name for dehydrated food waste other than garbage. Perhaps other contributors to this text will provide some of the answers.

References

Callis, J. J., P. D. McKercher, and M. S. Shahan. 1975. Foot-And-Mouth Disease. In: H. W. Dunne and A. D. Leman (Ed). Diseases of Swine. 4th Rev. Edition. p. 325. Iowa State University Press. Ames, IA.

Childs, T. 1952. The history of foot-and-mouth disease in Canada. Proc. 56th Ann. Meeting. U.S. Livestock Sanit. Assoc., p. 153.

CFR. 1998. Title 9 Code of Federal Regulations. Section 166 - Swine health protection; Section 167 - Rules of practice governing proceedings under the swine health protection act. U.S. Government Printing Office.

Congressional Quarterly Weekly Report Vol. 38 (Oct.-Dec.1980), p.2889-3680.

Congressional Record, vol.126, p. 2733-4176.

Dunne, H. W. 1975. Hog Cholera. In: H. W. Dunne and A. D. Leman (Ed). Diseases of Swine. 4th Rev. Edition. p. 189. Iowa State University Press. Ames, IA.

House, J.A. and C.A. House. 1992. Vesicular Diseases. In: A. D. Leman, B. E. Straw, W. L. Mengeling, S. D'Allaire, and D. J. Talyor (Ed). Diseases of Swine. 7th Rev. Edition. p. 387. Iowa State University Press. Ames, IA.

House Reports vol. 19, 13377, 1980. House of Representatives.

Law, J. 1915. History of foot-and-mouth disease. Cornell Vet 4:224.

Madin, S. H. 1975. Vesicular Exanthema. In: H. W. Dunne and A. D. Leman (Ed). Diseases of Swine. 4th Rev. Edition. p. 286. Iowa State University Press. Ames, IA.

Mauer, F. 1975. African Swine Fever. In: H. W. Dunne and A. D. Leman (Ed). Diseases of Swine. 4th Rev. Edition. P. 256. Iowa State University Press. Ames, IA.

Mexico-United States Commission for the Prevention of Foot-and-Mouth Disease 1972.

Mohler, J.R. 1929. The 1929 outbreak of foot-and-mouth disease in California. J.Am. Vet. Med. Assoc. 75:309.

Mohler, J. R. 1938. Foot and mouth disease. USDA Farmers' Bulletin. 666:1 (rev 1952).

Mohler, J.R. and Traum, J., 1942. Foot-and-mouth disease. Separate No. 1882, Keeping Livestock Healthy. Yearbook of Agriculture, USDA, p. 263.

Mulhern, F.J., 1953. Present status of vesicular exanthema eradication program. Proc. 57th Ann Meet. US. Livesto. Sanit Assoc. pp. 326-333.

Ribeiro, J.M. and R.J. Azevedo. 1961. Reapparition de la peste porcine (P.P.A.) au Portugal. Bull. Off. Intern. Epizoot. 55:88.

Ribeiro, J.M., R.J. Azevedo, M.J.O. Teixeiro, M.C. Braco Forte, R.Rodriguez, A.M. Ribeiro, E. Oliveiro, E., F. Noronha, C. Grave Pereira, and J. Dias Vigario, 1958. Peste porcine provoquee par une souche differente (souche L) de la souche classique. Bull. Off. Intern. Epizoot. 50:516.

Shahan, M.S. 1954. Present situation on foot-and-mouth disease. Mil. Surg 114:444.

USDA Release, 1947. Summary developments in the Mexican outbreak of foot-and-mouth disease. Jan 28, 1947.

USDA Release, 1953. No. 1279-53. May 28, 1953.

USDA Release, 1954a. No. 999-54, April 14, 1954.

USDA Release, 1954b. No. 329-54. Dec. 22, 1954.

Van Oirschot, J.T., 1992. Hog Cholera. In: A. D. Leman, B. E. Straw, W. L. Mengeling, S. D'Allaire, and D. J. Talyor (Ed). Diseases of Swine. 7th Rev. Edition. p. 274. Iowa State University Press. Ames.

6

Food Waste as Swine Feed

by Michael L. Westendorf

Introduction

Food waste is commonly fed to pigs in many parts of the United States, Puerto Rico, and around the world (Westendorf et al. 1996). Food plate waste comprises nearly 22 million tons, or 8.9%, of all Municipal Solid Waste (MSW). It causes a disproportionately large amount of disposal costs due to odors, gas production, and rodent control at landfills. Since food waste often has a very high nutritional value, it may be fed to live-stock. Texas, Florida, and New Jersey are the leading states in the dis-posal of food waste as livestock feed. This waste is composed primarily of food and table plate waste, vegetable and food-processing waste, bakery waste, and waste from dairy product and egg processing. The Swine Health Protection Act requires that all table waste fed to swine be cooked at boiling temperature (100°C) for 30 min prior to feeding. There are some states that have banned food waste use altogether. The cooking methods currently approved by the USDA-APHIS are either steam-cooking or by cooking over an open flame. The intention of the cooking requirement is to eliminate the risk of animal diseases such as hog cholera, foot-and-mouth disease, African swine fever, and swine vesicular disease, which infect swine. This requirement is also meant to reduce the risk that infectious organisms (*Salmonella, Campylobacter, Trichinella*, and *Toxoplasma*) are transferred to people.

Although the feeding of wet food waste to animals has been scruti-nized, there are still at least 30 states (Polanski 1995) that allow it in some form. Energetically, feeding wet food waste avoids the costs of dry-ing and processing that typify some of the newer processing methods (extrusion, pelletizing, and dehydration). Regardless of the processing method, cooking is still required to meet the requirements of the Swine Health Protection Act (U.S. Congress 1980). Cooking or heat process-ing is also required by the FDA to meet the requirements of the Ruminant Feeding Ban (Animal Proteins Prohibited in Ruminant Feed

69

1998). The feeding of wet food waste is useful for producers who are not willing to further process the waste, and if they can accept the reductions in animal performance associated with wet food waste feeding, they can attain an economic advantage due to its low cost.

Nutrient Composition

Table 1.4 in chapter 1 lists the nutritive value of various types of food waste fed in the 1960s. This represented food waste collected from hotels and restaurants, institutions such as nursing homes, hospitals, and prisons, military bases, and municipalities (Kornegay et al. 1970). Crude protein and fat (measured by ether extract) were adequate, if not abundant in most food wastes. The mineral and vitamin contents of food waste were adequate also when compared to traditional (i.e., corn and soybean meal) diets, although calcium, phosphorous, and pantothenic acid were either borderline or deficient in some food-waste sources. The high moisture content of the waste sampled (approximately 80% water) represents the chief limitation of feeding food waste because it often reduces feed intake and makes storage of the waste impossible. The digestibility of these same sources of food waste varied greatly (Kornegay et al. 1970), and these researchers concluded that pigs fed food waste performed adequately when supplemented properly, even when fed the poorest food-waste sources. However, pigs fed hotel and restaurant waste, military waste, and institutional waste all outperformed pigs fed municipal waste, possibly because of the higher fiber and ash in municipal waste that led to decreased feed intake. The addition of corn and soybean meal supplements to municipal waste resulted in pig performance approaching that of traditional corn/soy diets.

Westendorf et al. (1999) reported that food waste collected in recent years also had adequate or good nutrient content. Table 6.1 shows the results of a survey of 54 to 63 samples collected from swine farms that feed food waste. This food waste originated from casinos, restaurants, or institutions (nursing homes, hospitals, prisons, etc.). These farms supplemented food plate waste with bakery waste, fish waste, and/or various types of vegetable waste. Only food (plate) waste was sampled for analysis except for some fish waste added during cooking on one farm. The data in table 6.1 reveal that food waste has adequate protein (~21%) and high fat (~26%). Dry matter is low (~27%) as is fiber (~6%). Levels of minerals were generally adequate, although calcium was very high on one farm and zinc was borderline or low on several farms. Sodium averaged ~1% of dry matter, similar to that observed by Kornegay et al. (1970) and Myer et al. (1994).

Table 6.1. Average nutrient content of food plate waste fed on sample farms

Nutrient[a]	Sample Size	Mean	SD[b]	CV[c]	Range
DM (%)	63	27.00	5.20	19.3	13.0 to 39.6
CP (%)	63	20.80	5.70	27.5	13.6 to 37.7
EE (%)	63	26.30	8.00	30.4	9.1 to 46.9
ADF (%)	62	6.30	2.60	41.2	2.4 to 15.3
Ash (%)	63	6.20	2.20	35.3	3.0 to 16.4
Ca (%)	63	0.92	1.02	111.1	0.06 to 6.33
P (%)	63	0.64	0.46	72.1	0.12 to 2.18
Mg (%)	63	0.08	0.03	34.8	0.03 to 0.13
Na (%)	63	1.04	0.37	35.5	0.63 to 1.79
K (%)	63	0.83	0.43	51.6	0.13 to 2.01
Cu (mg/kg)	54	17.30	23.50	136.4	1.4 to 164.6
Fe (mg/kg)	63	441.00	314.00	71.0	78 to 1778
Zn (mg/kg)	63	63.00	201.00	321.0	10.6 to 1621
Mn (mg/kg)	54	21.00	15.60	74.4	5.7 to 58.4

[a]All nutrients reported on a dry matter basis.
[b,c]Standard deviation and coefficient of variation.
Source: Westendorf et al. 1999.

High levels of meat in discarded plate waste may have led to the high levels of protein observed (Westendorf et al. 1999), and meat fat trim may have led to the increase in fat percent. Increased sodium may relate to the use of seasonings in prepared food. Most essential amino acids were adequate except for lysine on one farm. (The samples used for amino acid analysis were composites of previous collections.)

The most important observation from this study was that food waste has adequate, if not excellent, nutritional profile. However, the high moisture content makes it more difficult to collect and feed, limits shelf life, and reduces dry matter intake. In addition, variable CVs, often in excess of 100%, tend to make food waste difficult to incorporate into commercial swine diets.

Feeding Value

The feeding of food waste or garbage is not new. In studies completed more than 30 years ago, Kornegay et al. (1970) analyzed several sources of food waste: hotel and restaurant, military, institutional, and municipal wastes. All of these sources of food waste were adequate if supplemented properly, with the exception of municipal waste, and even unsupplemented animals gained more than 0.45 kg/d (municipal waste was the exception). Feeding diets ranging in moisture content from 10 to 85%, Kornegay and Vander Noot (1968) found that pigs adapted to the greater moisture content of the diet with increases in dry matter intake

Table 6.2. Nutrient digestibility and daily nitrogen balance of a food
waste (FW) and a corn/soybean meal (CSM) diet

	FW	CSM	Standard Error
Intake	1360.00	1698.00	84.200
Nutrient			
DM (%)	86.69	85.60	0.480
CP (%)	88.16[a]	84.29[b]	0.800
ADF (%)	53.71[a]	70.58[b]	1.750
EE (%)	93.51[a]	27.62[b]	1.500
Ash (%)	77.09[a]	62.53[b]	0.930
NFE (%)	97.22[a]	93.77[b]	0.650
Nitrogen			
Intake (g)	44.05	53.19	2.680
Excretion (g)	19.51	23.93	2.060
Balance (g)	24.54[a]	29.26[b]	1.060
Retention (%)	55.96	55.22	2.100
Amino Acid			
Digestibility (%)	91.00[a]	86.60[b]	0.008

Source: Westendorf et al. 1998.
Note: Nutrient content of food waste averaged 22.0% dry matter,
20.2% crude protein, 15.2% acid detergent fiber, and 3.7% ash.
[a,b]Values in the same row were significantly different at $P < 0.05$.

and daily gains as adaptation time increased. Similarly, Kornegay et al.
(1970) reported that rats also adapted to food waste diets with compensatory dry matter intake and increased weight gains. Kornegay and
Vander Noot (1968) also reported that diet digestibility was unaffected
as the moisture content increased, a result supported by Kornegay et al.
(1970), who found that different types of food waste can have good
digestibility, despite high moisture content. However, although animals
may compensate and digestibility may be adequate, dry matter intake is
still the chief limiting factor to growth rates when feeding food waste.

In a study of food waste (22% dry matter) collected from a cafeteria
compared to a corn and soybean meal diet fed to growing pigs (35.4 kg),
Westendorf et al. (1998) reported that dry matter digestibility did not
differ between the two diets, while both crude protein and amino acid
digestibility were greater for the food-waste diet (see table 6.2). The
food waste used in this study was cooked prior to feeding to meet the
requirements of the Swine Health Protection Act (U.S. Congress 1980).
While nitrogen balance was lower in the group receiving food waste,
probably due to a lower dry matter intake, nitrogen retention, as a percentage of nitrogen intake, did not differ between the two groups. This
indicates that food waste can be an excellent protein source, probably

due to the presence of animal products that have a good mixture of essential amino acids. The research described in the previous section also found that food waste often has a good blend of essential amino acids.

Water intake, both dietary and total, was higher for the food waste-fed pigs than for those receiving corn and soybean meal. Food waste-fed pigs only drank 0.21 liters/day. Both urinary and total water excretion were greater in food waste-fed pigs, but daily water balance was similar between the two groups. Food waste provided a source of water for pigs, but it reduced dry matter intake substantially (food-waste pigs dry matter intake = 1.36 kg/day; corn and soybean meal pigs dry matter intake = 1.70 kg/day).

In another experiment, Westendorf et al. (1998) supplemented pigs fed cafeteria food waste with two levels of ground corn plus a vitamin/mineral premix. A corn and soybean meal diet was the control. Assuming that energy would be the limiting nutrient and that protein needs would be met by food waste, treatments were (1) a corn/soybean meal diet fed ad libitum; (2) pigs fed food waste ad libitum plus a supplement fed at 50% of Treatment 1 intake; (3) pigs fed food waste ad libitum plus a supplement fed at 25% of Treatment 1 intake; and (4) pigs fed food waste ad libitum. The supplements fed to Treatments 2 and 3 were identical except they contained differing amounts of premix in order to maintain a vitamin/mineral intake (not counting food waste) similar to Treatment 1. This experiment was repeated in both growing and finishing pigs. Results are presented in table 6.3. Pigs fed only food waste in this experiment gained more than 0.45 kg/day. During both the growing and finishing periods, pigs fed food waste ad libitum gained faster during the second 14 days of the experiment, indicating the pigs adapted with compensatory growth. In fact, the pigs fed food waste gained slightly faster than the control pigs fed corn and soybean meal during the second period of the finishing phase (0.79 kg/day v. 0.76 kg/day). Supplementation resulted in significantly (P < 0.05) improved performance over the ad libitum food waste group in both the growing and finishing trials. In the finishing trial, the group fed food waste plus supplement at 50% of the amount of the control group gained 0.90 kg/day, not significantly different from the control group.

Both Kornegay et al. (1970) and Westendorf et al. (1998) indicated that supplementation of food waste results in improved animal performance. While the crude protein (Westendorf et al. 1999) of food waste averaged 20.8% (table 6.1), seemingly well in excess of the needs for growing and finishing pigs, the high moisture content (27%) in food waste may limit intake and result in deficiencies. However, in the finish-

Table 6.3. Gain and feed intake of pig fed a food waste, corn/soybean meal, or food waste plus supplements

Growing Phase	1-CSM	2-FW + 50% CSM	3-FW + 25% CSM	4-FW	Standard Error
ADFI	1.96	1.68	1.30	0.915	—
ADG 0-14	0.74[a]	0.51[b]	0.44[b]	0.280[c]	0.48
15-28	0.90[a]	0.80[ab]	0.78[ab]	0.610[b]	0.54
0-28	0.82[a]	0.66[b]	0.61[b]	0.460[c]	0.43

Finishing Phase	1-CSM	2-FW + 50% CSM	3-FW + 25% CSM	4-FW	Standard Error
ADFI	2.77	2.69	2.02	1.360	—
ADG 0-14	1.26[a]	0.87[b]	0.77[b]	0.580[c]	0.34
15-28	0.76	0.97	0.80	0.790	0.86
29-42	0.94[a]	0.85[a]	0.73[ab]	0.490[b]	0.89
0-42	0.99[a]	0.90[a]	0.77[b]	0.620[c]	0.36

Source: Westendorf et al. 1998.
Treatments: 1, diet A, corn/soybean meal (CSM) fed ad libitum; 2, food waste (FW) fed ad libitum plus supplement fed at 50% of Treatment 1 intake; 3, FW fed ad libitum plus supplement fed at 25% of Treatment 1 intake; and, 4, FW fed ad libitum.
Note: Nutrient content of food waste averaged 22.4% dry matter, 21.4% crude protein, 14.1% acid detergent fiber, and 3.2% ash.
[a,b,c]Values with a letter in common are not significantly different, otherwise different at $P < 0.05$.

ing trial described above (Westendorf et al. 1998), because the crude protein was adequate in food waste, the pigs fed the high level of supplement (corn plus a vitamin/mineral premix) gained at a level similar to the control pigs. In some earlier supplementation studies (Kornegay et al. 1970), pigs responded to added crude protein in the supplements fed, possibly because the crude protein levels of food waste averaged less (16.1%) than the 21.4% in the study by Westendorf et al. (1998). This indicates that crude protein often may be adequate in the diets of growing and finishing pigs, but it also underscores the variability of food waste.

Although digestibility and nitrogen metabolism of food-waste diets have been good and often comparable to more traditional diets, the high moisture content of food waste decreased intake and led to slower growth rates. Two problems are commonly associated with feeding food waste. First, the high moisture content reduces the shelf life of food waste, reduces food intake of pigs, and is difficult to incorporate into any kind of complete diet. Second, the variable nutrient content of food waste makes balancing diets difficult. Although the average nutrient content of food waste is often good, the huge day-to-day variability (CVs for individual nutrients in excess of 100%, Westendorf et al. 1999) means that properly supplementing food waste diets will be difficult. When this is done, as in the studies reported by Kornegay et al. (1970) and Westendorf et al. (1998), supplemented food waste fed ad libitum results in acceptable performance. Of course, the cooking mandate (USDA-APHIS, 1990) is a time-consuming and inefficient process and another disadvantage to feeding food waste.

Feeding food waste wet is still the most common feeding method. A recent study by USDA-APHIS (1995a, b, c) indicated that 123,000 pigs were fed on more than 2,000 farms nationwide with each farm averaging 39,770 kg of food waste fed per year. New technologies should focus on removing moisture and reducing variation by blending with other more stable products such as wheat middlings or corn, in order to increase the use of food waste as an animal feed. With 21.9 million tons of food waste produced annually and only 2.4% of this recycled (Franklin and Associates 1998), there are huge opportunities for increased food-waste recycling.

Meat Quality

Previous research about the quality of meat from pigs fed food waste has been inconclusive. Some researchers (Lovatt et al. 1943; Modebe 1963) found softer fat in pigs fed food waste, while others (Hunter 1919; Engel et al. 1957) found that food waste did not affect iodine number or melting point. Engel et al. (1957) and Peterson (1967) found no differences between dressing percentage and loin-eye area in pigs fed food waste, food waste plus supplement, or concentrate diets. Kornegay et al. (1970) indicated some variability in carcass characteristics. One trial showed that dressing percentage and backfat thickness were reduced in pigs fed food waste more than for pigs fed food waste plus a supplement. In another trial, dressing percentage was higher when pigs were fed food waste alone than when pigs were fed food waste plus a supplement. Kornegay et al. (1970) also observed by manual evaluation that pigs fed food waste alone had softer fat than did pigs fed food waste plus supplement or concentrate.

Myer et al. (1999) compared the use of a dehydrated food-waste product added to the diet at 0, 40, or 80% (the product that was added was made by mixing 60% food waste and 40% concentrates, pelleting and drying). In two finishing trials, they found no differences ($P > 0.10$) in carcass lean content or other carcass characteristics. However, in trial 2, there was a linear decrease in backfat thickness ($P < 0.05$) and a linear increase in carcass fat softness ($P < 0.01$) as the percentage of

Table 6.4. Mean ratings for intensity and preference (liking) for pork taste of pigs fed either food waste (FW) or corn soybean/meal (CSM)

	Intensity			Preference		
Attributes	FW	CSM	Standard Error (SE)	FW	CSM	SE
Pork flavor	7.65[a]	6.61[b]	0.44	7.13	7.09	0.46
Chewiness	7.65[c]	8.97[d]	0.37	8.97[c]	7.26[d]	0.43
Juiciness	9.03[c]	6.49[d]	0.43	8.95[c]	6.82[d]	0.44
				Overall preference		
Flavor				7.21	6.70	0.46
Texture				8.46[a]	7.42[b]	0.44

Source: Westendorf et al. 1998.
Note: Values are reported as 1 to 15 on a 15-cm line scale.
[a,b]Values in the same row were significantly different at P = 0.05.
[c,d]Values in the same row were significantly different at P < 0.05.

dehydrated food waste in the diet increased from 0 to 80%. A fatty acid profile of the dehydrated food waste indicated the presence of a variety of products of both animal and vegetable origin. The data from this study supports previous projects that found increased softness of fat when food waste was fed.

A research trial by Westendorf et al. (1998) compared the taste of meat from pigs fed either food waste or a corn and soybean meal diet in a consumer taste panel. Twelve pigs (six fed food waste, six fed the corn and soybean meal diet) were fed to a finishing weight of 110 kg. The loins were frozen to use for palatability testing by 65 people participating in a consumer panel. Loin chops were thawed and cooked at 350°F until done. Cooked meat was transferred then to cutting boards, cut into 3.8 cm cubes, and offered to consumers. Each person tasted two pork meat samples, one from each experimental group, without knowledge of origin. Pork was evaluated for flavor, chewiness, and juiciness, and rated for both intensity and preference (also referred to as liking) using a scale from 1 to 15. Higher preference scores indicate a greater preference. Higher or lower intensity scores indicate the degree of intensity of the individual attribute (flavor, chewiness, and juiciness).

The results (see table 6.4) of the taste test indicated that the meat from food waste-fed pigs had acceptable organoleptic quality. The panel liked meat from pigs fed food waste and rated it juicier than meat from pigs fed the corn and soybean meal diet. The consumers rated meat from food waste-fed pigs as having a more intense flavor, but rated flavor preference equal between the two groups.

Most of the research done with food waste has shown carcass characteristics to be comparable to control animals, except for carcass fat, which has been softer in several research trials. Taste preference, as described in this current study, further supports the hypothesis that meat from pigs fed food waste is acceptable.

Processed Food Waste

There has been more research conducted recently using processed food waste as swine feed (Myer et al. 1994; Rivas et al. 1994; Altizio et al. 1998; Myer et al. 1999). There are a variety of processes that may be used for processing food waste, including dehydrating, pelleting, and/or extruding. According to Dr. Arnold Taft of the USDA-APHIS, processing methods still have to meet the requirements of the Swine Health Protection Act (U.S. Congress 1980). This means that food waste included as part of a blended, dehydrated, cooked, extruded, or pelleted product must be cooked at 100°C for 30 min. However, many of the processes utilized result in temperatures in excess of 100°C and should meet the spirit of the requirements for cooking. Processed food waste will be evaluated on a case-by-case basis and may be classified as a rendered product for regulatory purposes. These determinations will be made by state or USDA-APHIS veterinary services or by the FDA, depending on who has jurisdiction.

In numerous trials with a dehydrated restaurant waste blended with concentrate feeds, Myer et al. (1994, 1999) found good results at inclusion rates up to 80% of the blended product (48% food waste). The use of processed food waste shows good promise for increasing the utilization of food waste as animal feed. Some of the chief issues influencing new processing techniques are regulatory in nature, such as whether rendering food waste can substitute for the Swine Health Protection Act and bring regulatory authority to the FDA instead of USDA-APHIS. This chapter has focused on the established method of feeding food waste wet; the use of processed food waste as animal feed is discussed elsewhere in the book.

Risk Assessment

A 1995 survey of food-waste feeders by USDA-APHIS documented the practice of feeding recycled commodities to swine in the United States and sought to determine the degree of risk from feeding uncooked food waste. The survey identified the different types of food waste being fed,

Table 6.5. Sources and types of plate waste fed to swine

State	Source			
	Households	Jails	Schools	Restaurants
	(% of Total Fed[a])			
Puerto Rico	48.5	3.0	39.0	5.8
Texas	0.1	23.3	37.2	37.4
Florida	0.1	36.7	18.1	40.1
New Jersey	15.9	28.1	4.1	46.1
Hawaii	8.8	5.6	13.0	64.1
Minnesota	0.0	3.0	3.3	92.7
United States	2.4	22.1	23.3	45.9

State	Type		
	Plate Waste	Bakery Waste	Fruit and Vegetable Produce
	(% of Total Fed[b])		
Puerto Rico	93.7	1.7	2.4
Texas	90.6	5.3	2.7
Florida	92.0	2.5	4.3
New Jersey	65.1	23.2	4.8
Hawaii	81.5	2.8	14.3
Minnesota	74.6	12.2	13.1

Source: USDA-APHIS 1995a.
[a] Remainder made up of plate waste originating from grocery stores, hospitals, nursing homes, and the military.
[b] Remainder made up of animal products (fish, eggs, unpasteurized dairy products, carcasses, etc.) and miscellaneous.

the number of hogs fed, and the leading states. It also determined the relative risk from different types of food waste, such as institutional waste (hospital, prison, nursing home, etc.) v. municipal or household waste.

According to these results (USDA-APHIS, 1995a, b, c), 1.2 million pounds of food waste is fed daily in the lower 48 states, 140,000 pounds in Hawaii, and 420,000 pounds in Puerto Rico, more than any other area. The leading states for feeding food wastes are Texas, Florida, New Jersey, Hawaii, and Minnesota. Texas has the most feeders per state (excluding Puerto Rico), and New Jersey has the largest feeders.

The sources and types of food waste are compared in table 6.5. Puerto Rico feeds a greater percentage of household food waste than do any of the other states compared, and many states feed more than just plate waste, as table 6.5 shows. Texas, Florida, and Puerto Rico all feed more than 90% plate waste. New Jersey feeders used only about 65%. The remainder is often made up with bakery waste or fruit and vegetable waste. This part of the survey played an important part of the risk assessment because these different food-waste types present different risks. The nonplate food waste should be of no risk unless raw meat or carcasses are fed.

The diseases of concern in this risk assessment were several foreign diseases of animals (hog cholera, foot-and-mouth disease, African swine fever, and swine vesicular disease) and other pathogens having public health significance (*Salmonella, Campylobacter, Trichinella,* and *Toxoplasma*). Meat imports were monitored to determine the presence of both legal and contraband materials at ports of entry and whether these products would be in the swine food chain. This information was analyzed by country of origin, relative infection rate of the country of origin (foreign animal disease infection rate), and how these products might go through the food chain. For example, contraband pork is assumed to be discarded always as household waste, while legally imported but inadequately processed pork products were assumed to have come from other sources, such as restaurants and institutions. The relative risk was estimated based on the survey of food-waste feeders described above and their sources of food waste. Data are expressed in table 6.6 as median probability (%) that contaminated food waste will reach susceptible swine within a year. This does not include any effects from on-farm cooking or treatment. Table 6.6 presents overall risk for both the United States and Puerto Rico.

Table 6.6 Median probability that contaminated waste will reach susceptible swine within a one year period of time

	Hog Cholera	Foot-and-Mouth Disease	African Swine Fever	Swine Vesicular Disease
	Percent Risk from Contraband Products[a]			
United States	6.7	4.1	0.53	0.5
Puerto Rico	23.9	22.2	3.80	0.2
Hawaii	4.6	2.6	0.00	0.3
	Percent Risk from Legal, Improperly Processed Imports			
United States	1.2	0.4	0.03	0.5

Source: USDA-APHIS 1995a.
[a]Contraband defined as improperly cooked pork products entering the United States illegally.

Contraband materials pose the greatest risk for swine fed food waste and are more often associated with household waste than with institutional waste (USDA-APHIS 1995a). Many food waste feeders feed no or little household waste, and their risk of feeding contaminated material is reduced accordingly. The risk assessment concluded that cooking (U.S. Congress 1980) food wastes is still warranted, but perhaps not when other risks are low.

The risk of *Salmonella, Campylobacter, Trichinella,* or *Toxoplasma* being fed was also evaluated (USDA-APHIS 1995a). Except for *Trichinella,* the risk estimate was 100% that a food-waste feeder would feed waste contaminated with one of these organisms within one year. *Trichinella spiralis* has been associated with the consumption of undercooked pork. Regulations for cooking food waste will eliminate this risk factor. However, Schad et al. (1987) indicated that the risk of *T. spiralis* is associated more with the presence of rodents around the feeding area than from feeding food waste. In New Jersey (E. W. Zirkle, Personal Communication), it has been determined that effective rodent control can help to eliminate the risk of *T. spiralis.*

This risk assessment (USDA-APHIS 1995a, b, c) concluded that cooking of food waste is still needed. The Swine Health Protection Act (U.S. Congress 1980; Public Law 96-468) defined garbage as "all waste material derived in whole or in part from the meat of any animal (including fish and poultry) or other animal material, and other refuse of any character whatsoever that has been associated with any such material, resulting from the handling, preparation, cooking, or consumption of food, except that such term shall not include waste from ordinary household operations which is fed directly to swine on the same premises where such household is located." Following passage of this Act, USDA-APHIS (1982a, b) submitted the following regulations that detail cooking requirements:

a) Garbage shall be heated throughout at boiling (212°F or 100°C at sea level) for 30 min.

b) Garbage shall be agitated during cooking, except in the steam-cooking equipment, to ensure that the prescribed cooking temperature is maintained throughout the cooking container for the prescribed length of time. (9CFR Part 166, 47FR 49940-49948).

Table 6.7 compares food plate waste feeding and cooking requirements for some of the 50 states. Some states accept the federal standard for cooking food waste while the remainder have either made the regulation more stringent or outlawed the practice entirely.

Table 6.7. Individual state garbage-feeding laws

State	Is Garbage Feeding Allowed?	Garbage Definition	Treatment Required
California	Yes	animal waste	Boil at 100° C for 30 min
Colorado	Yes	animal, fruit, and vegetable waste excluding vegetable leaves and tops	Boil at 100° C for 30 min
Georgia	No	animal, fruit, and vegetable waste	N/A
Illinois	No	animal, poultry, fish, and vegetable waste, but not the contents of the bovine digestive tract	N/A
Indiana	Yes	solid and semi-solid animal and vegetable waste	Only rendered product can be fed
Iowa	No	animal, fruit, and vegetable waste, but not animal waste that is processed at slaughterhouses or rendering establishments and is heated at 100° C for 30 min	N/A
Kansas	Yes	animal, fruit, and vegetable waste	Boil at 100° C for 30 min
Minnesota	Yes	animal, fruit, and vegetable refuse, but not waste from canned or frozen vegetables	Boil at 100° C for 30 min
Missouri	Yes	animal, fruit, and vegetable waste	Boil at 100° C for 30 min
Nebraska	No	fruit, vegetable, meat, and poultry material, but not nonmeat by-products from commercial food processors	N/A

(continued)

Table 6.7. Individual state garbage-feeding laws (*continued*)

State	Is Garbage Feeding Allowed?	Garbage Definition	Treatment Required
Nevada	Yes	animal material or other refuse associated with animal material	Left to the State Board of Agriculture and the State Quarantine Officer
New Jersey	Yes	animal waste and other refuse that has been associated with animal waste	Boil at 100° C for 30 min
New York	No	animal and poultry waste	N/A
North Carolina	Yes	animal waste	Boil at 100° C for 30 min
North Dakota	Yes	animal and vegetable waste, but not dairy products from a licensed creamery or dairy	Boil at 100° C for 30 min
Ohio	Yes	animal, fruit, and vegetable waste	Boil at 100° C for 30 min
Pennsylvania	Yes	animal and vegetable waste	Boil at 100° C for 30 min
Vermont	Yes	animal and vegetable waste	Boil at 100° C for 30 min
Wisconsin	No	animal or vegetable waste containing animal parts	N/A

Source: Polanski 1995 and the American Society of Agricultural Engineers.
Notes: Overview of individual state laws, specific requirements will vary by state.
Animal waste is defined as any edible by-product from the slaughter, processing, or cooking of livestock carcasses including meat, bone, and organ tissue (i.e., liver).
100° C = 212° F.   N/A means garbage feeding is not allowed.

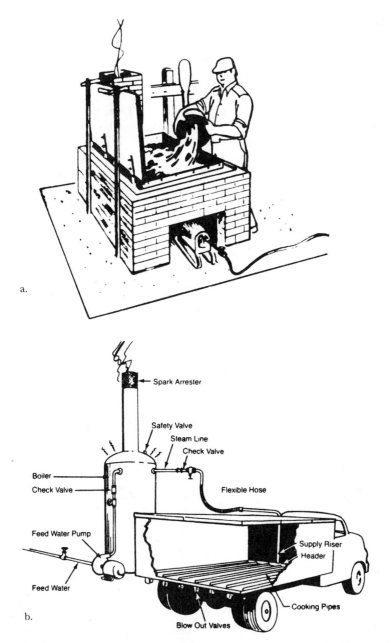

Figure 6.1. 6.1a. Diagram of food waste cooking over an open flame. 6.1b. Diagram of truck equipped for injecting steam for cooking.

Contemporary Practices

As described above and in other parts of this book, there is interest today in processing food waste to present a drier product for feeding, because dry products have longer shelf life, are easier to feed, and could be included as part of a complete diet. Nevertheless, the collection and feeding of wet food waste to pigs will likely continue for the foreseeable future. Tipping fees (Derr 1991), previously unmentioned, are the fees that food-waste generators (such as hospitals, institutions, restaurants, etc.) must pay to dispose of the waste. This could be at a landfill, incinerator, or compost site, but in the case of food waste, generators pay food-waste feeders to collect food waste and dispose of it by feeding it to animals.

The survey described in table 6.5 determined what types of food waste are being fed on farms feeding food waste to pigs. The plate waste all must be cooked, as described by the terms of the Swine Health Protection Act (U.S. Congress 1980). This cooking is done either over an open flame as shown in figure 6.1a or with steam injected into the food waste as shown in Figure 6.1b. Open flame cookers generally use either fuel oil or wood as material for burning, while steam cookers use fuel oil to boil water and produce steam that is injected into a container of food waste to bring it to boiling (USDA-APHIS 1990). Some food-waste feeders believe that cooking food waste will help to better mix the product while cooking. Figures 6.2 and 6.3 show pictures of an open

Figure 6.2. Open-flame cooking of food waste on a New Jersey farm. Fuel oil is used for cooking. Other operations may use wood or other flammable materials.

Figure 6.3. Feed wagon equipped for injecting steam into a load of food waste for cooking.

flame and a steam injection cooker, respectively. Veterinary inspectors visit food-waste feeders regularly to determine if cooking practices are working effectively. In New Jersey, the state veterinarian's office has primary enforcement responsibilities, but in many other states, the local USDA-APHIS office oversees enforcement.

Conclusion

In many ways, the management practices associated with the feeding of food waste have changed very little since this practice was first documented earlier this century (Minkler 1914; NJAES 1919). Cooking is now required as mandated by the Swine Health Protection Act (U.S. Congress 1980), although some states already required cooking at that time. The chief problems with food waste are its variability in nutrient content, high moisture content, and concerns about animal health and/or zoonoses. Cooking was required to reduce the animal health risks associated with feeding food waste. New processing practices that reduce moisture content will make it easier to feed food waste, yield a longer shelf life, and make it easier to incorporate food waste into commercial diets. The variability of food waste, as evidenced by extremely high CVs for most nutrients, indicates that even if food waste is fed dry, its variability will make it difficult to include food waste in today's highly programmed swine diets. Cooking of food waste should continue to be

Figure 6.4. Bakery waste stored in a commodity shed on a New Jersey swine farm. Nearly all New Jersey swine-feeders feed some form of bakery waste or by-product.

required whenever food waste is fed wet. The 100°C for 30 min cooking requirement was originally required to ensure uniform and thorough cooking throughout. This was enacted not only because of risks associated with feeding food waste, but also because of concerns that cooking varies from farm to farm (over an open flame or using steam). However, this requirement should be modified for new processes such as pelleting, dehydrating, and/or extruding, because these new technologies will mix and heat the product more efficiently, while blending with other feeds, and should result in a more uniform, safe product when cooked. (Figure 6.4 shows bakery by-product to be fed to pigs; Figure 6.5 shows food waste collected from a prison being mixed with dry feed and extruded for use as a dairy feed).

New technologies for processing must focus not only on pathogen reduction through heat treatment, but also on reducing the moisture content and in reducing nutrient variability. This could be accomplished by stricter controls on the sources of food waste used and/or by blending with other materials during processing. The very low (2.4%) recycling rate for food waste is one of the lowest for all municipal solid waste, despite the fact that food waste has excellent nutritional content. If the problems discussed here can be addressed, food waste could make a welcome addition as a feedstuff for livestock.

Figure 6.5. An example of food waste being extruded. Extrusion is one of several new technologies that holds promise for processing food waste into a drier product.

References

Altizio, B. A., P. A. Schoknecht, and M. L. Westendorf. 1998. Growing swine prefer a corn/soybean diet over dry, processed food waste. *J. Anim. Sci.* 76 (Suppl.1):185. (Abstr.)

Animal Proteins Prohibited in Ruminant Feed. 1998. Title 21 Code of Federal Regulations '589.2000. U.S. Government Printing Office.

Derr, D.A. 1991. Expanding New Jersey's Swine Industry Through Food Waste Recovery. Proceedings of Getting the Most From Our Materials Conference: Making New Jersey the State-of-the-Art. L. Gilbert and B. Salas, ed. New Brunswick, NJ.

Engel, R. W., C. C. Brooks, C. Y. Kramer, D. F. Watson, and W. B. Bell. 1957. The composition and feeding value of cooked garbage for swine. Va. Agr. Exp. Sta. Tech. Bul. 133.

Franklin and Associates. 1998. Characterization of Municipal Solid Waste in the United States: 1997 Update. U.S. Environmental Protection Agency. Municipal and Industrial Solid Waste Division. Office of Solid Waste. Report No. EPA530-R-98-007. Prairie Village, KS: Franklin and Associates.

Hunter, J. M. 1919. II. Garbage as a hog feed. Fortieth Annual Report of the New Jersey Agric. Exp. Sta. Trenton, NJ.

Kornegay, E. T. and G. W. Vander Noot. 1968. Performance, digestibility of diet constituents and N-retention of swine fed diets with added water. *J. Anim. Sci.* 27:1307-1311.

Kornegay, E. T., G. W. Vander Noot, K. M. Barth, G. Graber, W. S. MacGrath, R. L. Gilbreath, and F. J. Bielk. 1970. Nutritive evaluation of garbage as a feed for swine. Bull. No. 829. College of Agric. Environmental Sci. New Jersey Agric. Exp. Sta. Rutgers, The State Univ. of New Jersey. New Brunswick, NJ.

Lovatt, J., A. N. Worden, J. Pickup, and C. E. Brett. 1943. The fattening of pigs on swill alone: a municipal enterprise. *Empire J. Exp. Agr.* 11:182.

Minkler, F. C. 1914. Hog cholera and swine production. Circular No. 40. New Jersey Agric. Exp. Sta. Trenton, NJ.

Modebe, A. N. A. 1963. The value of African-type swill for pig feeding. J. W. African Sci. Assn. 8:33.

Myer, R. O., J. H. Brendemuhl, and D. D. Johnson. 1999. Evaluation of dehydrated restaurant food waste products as feedstuffs for finishing pigs. *J. Anim. Sci.* 77:685.

Myer, R. O., T. A. DeBusk, J. H. Brendemuhl, and M. E. Rivas. 1994. Initial assessment of dehydrated edible restaurant waste (DERW) as a potential feedstuff for swine. Res. Rep. Al-1994-2. College of Agric. Florida Agric. Exp. Sta. Univ. of Florida. Gainesville, FL.

NJAES. 1919. II. Garbage as a Hog Feed. Fortieth Annual Report of the New Jersey Agric. Exp. Sta. Trenton, NJ.

Peterson, L. A. 1967. Growth and carcass comparisons of swine fed a concentrate ration, cooked garbage, and additional protein, vitamin and mineral supplements. M.S. Thesis. Univ. of Conn., Storrs.

Polanski, J. 1995. Legalizing the Feeding of Nonmeat Food Wastes to Livestock. *Appl. Engr. Agr.* 11(1):115.

Rivas, M. E., J. H. Brendemuhl, D. D. Johnson, and R. O. Myer. 1994. Digestibility by Swine and Microbiological Assessment of Dehydrated Edible Restaurant Waste. Res. Rep. Al-1994-3. College of Agriculture. Florida Agric. Exp. Sta. Univ. of Florida. Gainesville, FL.

Schad, G. A., C. H. Duffy, D. A. Leiby, K. D. Murrell, and E. W. Zirkle. 1987. Trichinella Spiralis in an Agricultural Ecosystem: Transmission under Natural and Experimentally Modified On-Farm Conditions. *J. Parasit.* 73(1):95.

U.S. Congress. 1980. Swine Health Protection Act. Public Law 96-468.

USDA-APHIS, VS. 1982a. Swine Health Protection Provisions. USDA Animal and Plant Health Inspection Service. Federal Register. 47(213):49940-49948.

USDA-APHIS, VS. 1982b. State Status Regarding Enforcement of the Swine Health Protection Act. USDA Animal and Plant Health Inspection Service. Federal Register. 47(251):58217-58218.

USDA-APHIS, VS. 1990. Heat-Treating Food Waste—Equipment and Methods. USDA Animal and Plant Health Inspection Service, Veterinary Services. Program Aid No. 1324.

USDA-APHIS, VS. 1995a. Risk Assessment of the Practice of Feeding Recycled Commodities to Domesticated Swine in the U.S. United States Department of Agriculture Animal and Plant Health Inspection Service, Veterinary Services. Centers for Epidemiology and Animal Health. Fort Collins, CO.

USDA-APHIS, VS. 1995b. Swine waste feeder profile reveals types, sources, amounts, and risks of waste fed. United States Department of Agriculture Animal and Plant Health Inspection Service, Veterinary Services. Fort Collins, CO.

USDA-APHIS, VS. 1995c. Risk of feeding food waste to swine: Public health diseases. United States Department of Agriculture Animal and Plant Health Inspection Service, Veterinary Services. Fort Collins, CO.

Westendorf, M. L., E. W. Zirkle, and R. Gordon. 1996. Feeding food or table waste to livestock. *Prof. Anim. Sci.* 12(3):129-137.

Westendorf, M, L., Z. C. Dong, and P. A. Schoknecht. 1998. Recycled cafeteria food waste as a feed for swine: nutrient content, digestibility, growth, and meat quality. *J. Anim. Sci.* 76:3250.

Westendorf, M. L., T. Schuler, and E. W. Zirkle. 1999. Nutritional quality of recycled food plate waste in diets fed to swine. *Prof. Anim. Sci.* 15(2):106-111.

7

The Economics of Feeding Processed Food Waste to Swine

by Felix J. Spinelli and Barbara Corso

Introduction

Channeling food waste to swine feeders can create direct social benefits by providing an alternative outlet for food waste while at the same time, converting it into safe, nutritious pork products. Food waste-based swine rations have been used for centuries. Food-waste feeding to swine is often found in areas that have abundant sources of useable food waste, have available labor to handle it, do not produce sufficient feed grain supplies, and have food-waste generators facing limited landfill space.[1] Several new developments, such as the emergence of large-scale food processors and sales outlets (generating large quantities of useable food waste) and advances in new food-waste processing technologies, may increase interest in feeding food waste. These new supply-side developments may make food waste more available at specific locations for use or processing, thereby reducing the inconvenience in its collection and handling and relieving concerns associated with any health and food safety risks. This analysis shows that while current feeding operations incur all the costs in acquiring and using food waste (leaving them with small net benefits), most of the net benefits accrue to society. This apparently small net producer gain,[2] plus its current "inconveniences," help explain its limited interest in most areas of the United States. This situation probably will persist until landfills can no longer physically accept any additional food waste and/or technology makes food waste feeding more "user friendly." Given these two conditions, food-waste diversion from landfills to alternative uses could produce large social benefits. With swine's history in efficiently converting these materials, the developments could lead to common and widespread incorporation of food waste in swine rations.

The Economics of Feeding
Processed Food Waste to Swine

Many people have a negative perception of swine operations feeding processed food waste (PFW). These people fail to realize that food waste-based swine rations have been used for centuries (Van Loon 1988).[3] However, they may have seen one of the many marginal feeding operations that collect food waste from local schools, restaurants, and other small food-waste generators, treat it,[4] and feed it to a small group of swine in a "backyard" operation. In fact, poorly run operations spurred federal lawmakers to draft the original legislation in 1980, imposing minimal on-farm cooking times and temperatures for food waste fed to swine (CFR 1980). Although large-scale, well-managed PFW feeding operations exist, they are out-numbered by many poorly-run ones. Many of these poorly-run farms will persist as long as local authorities allow producers to (1) neglect swine health and animal welfare concerns and (2) dispose of their waste without regard to the local environment. On many of these operations, revenue from garbage pickup is the main source of income and hogs are simply viewed as "waste disposal machines on legs." These types of operations have the potential to become unsightly and environmentally unfriendly if hog waste and nonedible garbage contents are allowed to exceed "more-than-nuisance" proportions. Due largely to these local concerns, 17 states prohibit the feeding of food waste to swine[5].

The objective of this chapter is to make the case that, under certain conditions, garbage feeding of swine makes good economic sense. There are many current situations where garbage-fed swine are well cared for, receive nutritionally-sound diets comparable to conventional hog feeds, and produce a safe pork product while reducing the amount of food waste being sent to local landfills. Their success implies that PFW-feeding operations can be profitable while conforming to good agricultural production practices. These feeding operations view food waste as their main source of feed stock and appropriately refer to it as recycled or processed food waste. This chapter is broken into four parts. The first part describes the setting of PFW feeding in the United States in the early 1990s and factors that affected its use.[6] The second part discusses the technical issues involved in constructing the benefit and cost framework used in this analysis. The third section applies this platform to the statewide PFW-feeding situations in New Jersey and Florida and for the United States as a whole. A final section highlights the study's findings and draws some implications for the future.

The Current Setting and Some New Realities

In the early 1990s, the U.S. swine industry consisted of 235,840 commercial operations with roughly 6.5 to 7 million head in reproductive stock and around 50 million head being finished[7] at any one time (USDA—Hogs and Pigs Report 1995-1999).[8] Producers' annual marketings of roughly 100 million head of slaughter hogs generate $10 billion of gross farm income (USDA—Agricultural Statistics 1995-1999; USDA—Meat Animals 1995-1999) and supply the raw material to U.S. pork processors for annual domestic and foreign pork product shipments totaling over $15 billion (U.S. Department of Commerce 1996).

The majority of U.S. hog production takes place on highly capitalized, large-scale operations in the North Central region (which is composed of the Corn Belt, Lake States, and Northern Plains), but production is shifting to other areas, such as the southeast United States. For example, the leading states with respect to hog inventories (as a percent of the United States in 1993 and 1998) were Iowa (25.6 in 1993/23.4 in 1998), Illinois (10.0/7.8), North Carolina (8.5/15.8), Minnesota (8.4/9.5), Indiana (7.5/6.7), and Nebraska (7.4/5.7). Still, some hog production is reported in every state in many different types of facilities and under many different feeding situations (Shapouri et al. 1994). Even on the highly capitalized systems going into many of these new areas, feed still represents a major production cost. Most hog operations feed a standard finishing ration consisting of 80% feed grains, 15% soybean meal, and 5% mineral and nutritional supplement ingredients, but rations vary depending on producer preferences, cost, and local availability of substitute ingredients, particularly across feed grains, i.e., corn, small grains (Lawrence et al. 1988). Computer-assisted least-cost ration formulas to balance nutritional requirements with feasible least-cost feed sources have been used by the industry for more than 35 years.

One possible alternative feed source for swine producers is PFW, fed separately with supplements or as a component in mixed feeds. Presently, PFW-feeding to swine in the United States is of minor importance due to many reasons. These reasons involve either the farm-level demand or supply of PFW or both (See table 7.1). Farm-level demand factors include (1) lower relative feed efficiency that lengthens the time the hog is "on feed," often requiring additional feed additives, (2) producer reluctance related to on-farm handling problems and incorporating PFW into common production practices, and (3) price discounts on PFW-fed hogs sold at heavier-than-normal weights. Farm-level PFW supply factors include (1) quantity concerns, such as local and/or

Table 7.1. Factors affecting PFW-feeding to swine in the United States

Farm-level Demand	Farm-level Supply
Lower relative feeding efficiency	Inadequate availability of PFW
Lengthens time on feed/in bldgs	Seasonally
Requires add'l feed additives	Regionally
Producer reluctance	Quality inconsistency
Feed handling	Nutritional value/food safety
Farm prod. practices, i.e. cropping	Concerns for the potential of animal
patterns	disease/public health disease
Price discounts	
Penalty on price of fat hogs	

Affecting Both Demand and Supply of PFW
Restrictive federal and state regulations affect both farm-level demand and supply of PFW
High boiling temperatures/times may discourage innovation.

seasonal unavailability of PFW and (2) quality concerns, such as the amount of nonedible material in the original waste product received, often times its inconsistent nutritional value and its potential for animal and human disease transmission. Restrictive federal and state regulations transcend both farm-level demand and supply. For example, swine feeding of PFW is prohibited in 17 states, but where permitted, they must be licensed and are frequently inspected. Greatest numbers of waste feeders are found in Texas (871 licensed waste feeders), Hawaii (304), Florida (309), Arkansas (248), North Carolina (178), and New Jersey (31). In many of these areas, waste feeders are located close to large metropolitan areas assuring dependable amounts of waste food (USDA 1995c).

By any measure, past and current levels of PFW-feeding to swine in the United States are minuscule compared with total amounts fed or hogs produced. First, in relation to the annual number of swine fed in the United States, PFW-fed hogs represented only about 0.3% of slaughter (300,000 head of almost 93.1 million slaughtered in 1993). Second, the number of producers feeding PFW represents only about 1% of the total number of swine producers (2,283 operations out of 235,840). Finally, PFW makes up only 1.1% of the annual amount of feed fed to swine (550,000 tons of about 50 million tons fed to swine). Significant numbers of PFW-fed hogs are found in Texas (38,500), New Jersey (20,000), Missouri (17,350), Florida (15,500), North Carolina (7,800), and Oklahoma (4,000). Dividing hog numbers by the number of PFW-feeding operations given above suggests that the average size of operation varies greatly across states with 44 hogs per operation in Texas, 645 in New Jersey, 50 in Florida, and 44 in North Carolina.

Several market developments may increase interest in PFW-feeding to swine. First, the structure of the food-processing and distribution industries and final food sales outlets has become more concentrated (fewer firms, but each having larger operations) whereby concentrating the availability of relatively "pure" food waste. This greater concentration, both in the preparation and manufacture of many food products and in final retail food outlets (in such settings as restaurants, hotels, and other institutions) has the potential to provide the quantity and quality assurances needed to encourage more widespread use of PFW. For example, several institutional arrangements involving large-scale prepared meal manufacturers and/or large food users (such as schools, hospitals, prisons, theme parks, etc.) and PFW users could come together to guarantee a safe and stable flow of PFW to nearby waste feeders, collection sites, or further processing sites using new technologies. Secondly, new technological means are evolving to process food waste into more storable, safe PFW feed products. For example, food extruder and drier technologies could provide a product that could overcome many of PFW's current farm-level obstacles of greater acceptance. Such a food-waste recycling plan has been envisioned in Walt Disney World in Florida that would supply almost 70 tons of food waste per day for PFW processing (Acor 1994). Probably a more important economic driver than either of these forces is rising landfill costs. Faced with rising landfill charges, many of the food-processing and distribution institutions mentioned above may find it economical to divert their food waste from landfills to lower cost alternative outlets. These three developments — growing concentration of stable and safe PFW sources, new processing technologies, and rising landfill costs — point to the need to assess more closely any possible impediments, such as institutional, social, economic, regulatory, and others, that may stand in the way of increased use of PFW-feeding to swine.

The Technical Rationale Behind the Cost/Benefit Framework

Benefit/cost (BC) analysis is a practical way of assessing the desirability of projects, especially in situations that (1) bestow mostly society-wide benefits, (2) include considerable up-front investment costs, and (3) have long-term planning horizons (Prest and Turvey 1965). This section establishes the technical rationale concerning the underlying assumptions of each benefit and cost explicitly considered in this study.

Estimates of Benefits

Solid Waste Disposal Savings. The main benefit derived from feeding PFW to swine is the implied value of *saved* landfill capacity resulting from diversion of food waste from landfills to swine operations.[9] PFW feeders receive a payment from food-waste generators, but competition among food-waste collectors should drive these charges down to their true costs of collection and handling.[10] Given this assumption, landfill charges can act as a proxy for benefits derived by society-at-large. Total benefits are simply a function of the landfill tipping charge[11] multiplied by the amounts of PFW diverted from landfills. These fees were found to vary from less than $30 per ton to more than $120 per ton across the United States (Northeast Midwest Economic Review 1995). Fewer tons of solid waste arriving at landfills lower the rate at which this scarce resource is used up to the betterment of society-at-large. These gross savings could overstate social benefits by not accounting for any costs to separate and handle nonedibles in food waste before its use as PFW and their eventual disposal at a landfill.[12] No data were found on average nonedible content, therefore, no adjustment was made to gross savings.

A related important assumption in this area concerns the relative collection costs for garbage destined for a landfill or a PFW-feeding site. This study assumes that these costs are the same. This assumption simplifies this aspect of the analysis and is particularly helpful in dealing with solid waste removal costs because (1) solid waste collection charges vary greatly by location and type of garbage and (2) is highly confidential in nature, given the competitive nature of this industry. Thus, no net increase in solid waste collection charges is made when moving food waste to PFW-feeding operations, as opposed to landfill sites.

Producer Feed Cost Savings. The other significant benefit of feeding PFW is that producers need and purchase *less* commercial swine feeds. The total value of net producer feed cost savings depends on assumed per hog feed needs, their feed ration(s) and quality aspects, number of hogs fed, and ingredient prices[13]. Current conventional feeding programs are based on feeding about 1,000 pounds of feed (of about 80% corn/15% soybean meal) over 4 to 5 months to pigs initially weighing 45 pounds to a slaughter weight of 235 to 240 pounds. PFW-feeding programs vary greatly due to wide differences in feed values of PFW products. Derr reports that instead of a rate of gain of 1 pound of pork for 3 to 4 pounds of feed, the conversion rate for waste feeders can approach 1 to 21 pounds (Derr 1991a). Lower feeding value of PFW implies that prices, lower than conventional feeds, are required for economically rational producers to consider feeding PFW. For example, PFW at high feeding values, even in the range of 6 to 8 pounds of PFW to 1 pound of gain, implies that PFW must be priced one-half that of conventional feeds to remain competitive[14]. The impact of lower feeding equivalents

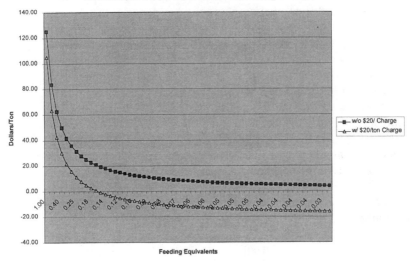

Figure 7.1. Shows the demand for processed food waste with and without handling charges.

on the demand for PFW can be seen in figure 7.1, which shows that if the price of feed is $125 per ton and the feeding value of PFW is 0.10 (10 pounds of PFW equals 1 pound of conventional feed), swine producers would only be willing to pay, or incur acquisition costs of, $12.50 per ton for PFW.

Estimates of Costs

Net Additional Handling Costs. The handling of PFW requires the movement of slurry-like and semi-solid food material from their source(s) to swine operations. Special equipment for hauling and storing, such as metal containers, and for cooking is required. Several large-scale operations have fairly sophisticated equipment — ranging from specially outfitted cement mixer-type vehicles for pickup, mixing, and cooking. Such equipment must be considered specific to the PFW-feeding operation and is probably more costly than conventional feeding apparatus. However, much of the same equipment, particularly the transport equipment, would be in service if this same food waste is collected and sent to a landfill. Also, the number of PFW-feeding operations with very low-cost garbage retrieval systems probably easily offsets the ones with the newer and more costly high-tech equipment. In addition, consider that most PFW-feeders avoid much of the cost associated with feeding systems based on conventional feeds. The above considerations, the offsetting cost structures of the few large (high-cost) operations against the many small (low-cost) ones and the avoidance of the costs associated with conventional feed systems, plus the lack of any cost data comparing the two, precludes raising or lowering the handling costs for PFW-feeders. Some claim that many PFW operators travel further and have high pick-up costs, but no data could be found to support this. One cost related to PFW-handling, the cooking requirement of 30 minutes at 212° F, deserves special mention. Fuel costs related to treating food waste varies depending on the equipment used (from simple open burning steel containers to steam-injected trucks), fuel type used (wood, butane, electricity), and the type of food waste treated. A PFW-cooking cost of $5 per ton is used throughout this study.

PFW-Feeding Inefficiencies. A fairly large body of research suggests that feeding PFW requires hogs to be "on feed" for an *additional two months* as compared with hogs fed a conventional feed ration. This same litera-ture shows that, on a strictly feeding value basis, most PFW must be priced at least one-half the cost of conventional feed to remain compet-itive. That calculation did not consider any additional time that hogs were on feed. The longer a hog is on feed, beyond what is absolutely necessary, puts highly capitalized operations at a disadvantage to use PFW. Longer time in the hog house denies space for other income-producing hogs. In economic jargon, this is called the physical oppor-tunity cost (POC) or the revenue foregone by the producer because the current stock of hogs is taking up the space that the next batch of hogs could be occupying and earning income. In effect, time is money because POC represents the physical cost of building space used by hogs. On many small operations with idle space and little interest in expanding production, the POC may approach zero. On highly capital-ized operations where space is an effective constraint on production "through-put," POC may approach the marginal value of product that comes from that unit of production (or facility). As a proxy, a rental charge on swine buildings can approximate the POC. Such a rental charge is calculated regularly and published by the USDA (USDA 1995b).

An additional cost in this area is the financial opportunity cost (FOC) of resources used in production. The FOC can be explicitly incurred through financial charges on capital tied up in buildings and equipment or implicitly incurred through the foregone interest on capital tied up in currently owned buildings. In effect, one must charge each batch of hogs for the amount of investment they represent. Producers either incur such costs directly in interest charges or indirectly on foregone interest not earned on committed capital. On many small operations with run-down (or in other words, fully depreciated) buildings, the FOC may approach zero. On larger operations with high fixed costs in buildings, the FOC may approach a full cost estimate of the (implicit or explicit) financial cost of buildings and equipment used in production. Any FOC estimate, obviously, would vary depending on the type of buildings and equipment used in production. For example, the average annual interest cost on real estate secured loans was $6,096 per livestock farm in 1992 (U.S. Department of Commerce 1992). Assuming that this interest cost is applicable to PFW-feeders and that they have a throughput of 2,624 pigs per year, the annual interest charge would be $2.32 per pig.[15] Dividing this charge by the average pig's life span on the farm (6 months) gives a monthly charge of about $0.39, or half of that if one assumes that two groups of pigs can move through the facility in a year. Obviously, this estimate is probably an under-estimation. It assumes a high turnover of stock and only considers the interest charge on *borrowed* capital. The FOC of buildings and equipment in swine production used in this study was found in USDA budgets (USDA 1995b). Such charges are particularly relevant in the event that additional buildings or equipment would be needed to accommodate increased PFW-feeding.

Relative Output Prices. There appears to be anecdotal evidence that many garbage feeders acquire "heavier than customary" feeder pigs (weighing about 90 pounds instead of 45 to 60 pounds) and feed them out to heavy weights at slaughter (290 pounds instead of 235 to 240 pounds). However, some top PFW-feeders using high-quality PFW apparently feed out their hogs much the same as conventional hog-feeding operations. These operations are assumed to be the exception and not the rule in PFW-feeding. Thus, most PFW-fed hogs are assumed to be marketed as "fat" hogs going into the lower-priced sow market, for such uses as sausage-making. These markets are typically priced anywhere from 15 to 35% lower than the lighter barrow/gilt market prices (USDA Agricultural Statistics 1994).[16] In 1993, this price discount was $8.26 per hundred weight or 18%. The PFW-fed hogs price discount used in this study is especially relevant if the added revenue per 300-pound hog does not cover the additional feed costs to get the hogs up to 300 lbs. Feeding trials have found that feed-to-meat conversion drops with heavier animal weights, as increased amounts of nutrients are diverted to sustain body functions. These feeding trial results seem to validate the common practice of feeding out hogs to a finishing weight of 220 to 235 lbs. Thus, it is assumed that failure to replace older, heavier hogs by younger ones decreases the total amount of meat produced from a given amount of feed supplied, resulting in a penalty to PFW-feeders.[17]

Relative Disease Risk-based Factors of Production. PFW-feeders may incur greater costs to safeguard against disease risk than producers using conventional feeds. These costs would fall into the following: (1) costs associated with any sudden and/or sustained lower production resulting from any diseases introduced as a result of PFW-feeding, (2) costs associated with any special sanitation precautions needed with PFW-feeding, and (3) any licensing and inspection costs related to PFW-feeding operations.

(1) Production Effect. It is assumed that our nation's animal or public health is not threatened as a result of current PFW-feeding[18]. Current regulations were designed to kill specific pathogens of swine, most notably swine vesicular disease and hog cholera. However, current conditions on many PFW-feeding operations point to higher pig mortality rates. This may be more attributable to typically poorer management practices found on many PFW-feeding operations (as compared with swine producers using conventional feeds), rather than PFW-feeding. An industry-wide mortality estimate for young pigs is around 3% (Purdue University 1976). This estimate is assumed to be double for PFW-feeders (an increase to 6%).

(2) Sanitation Precautions. While no direct link between food-waste feeding and increased disease potential has been established, poor sanitation found on many garbage- feeding operations, such as poor rodent control and flies, has been identified as a link to past disease outbreaks in several instances. On many PFW-feeding operations, especially small, under-capitalized ones, one often finds situations that are more predisposed to disease than on comparable conventional feeding operations. Some characteristics of these PFW-feeders include higher than average rodent populations, purchases of feeder pigs as a source of livestock (which may introduce disease into the herd), and numbers of free-range roving pigs in close proximity to wildlife populations. Of course, not all PFW-feeders have these problems. Regardless, it is assumed that PFW-feeders take additional safeguards to prevent disease outbreaks on their premises. These safeguards may include rat baits to monitor and control nearby rodent populations and cement flooring in feeding areas to facilitate cleaning and disposing of leftover PFW. A modest $100 per operation cost outlay for such items is assumed throughout this study.

(3) Licensing and Inspection Costs. Given current regulations, garbage-feeders are licensed annually and inspected at least quarterly to ascer-

tain if cooking equipment is capable of reaching the desired cooking temperatures and times. Licensing fees are minimal in most states, averaging about $15 per year (USDA-APHIS Internal Data). Due to their low costs, licensing fees are not included in this study. Also, no cost estimate on public time involved in filling out license application forms or in assisting with site inspections is included in this study.

Inspection costs are assumed to be borne by APHIS and state governments in equal proportions. Besides the inspection of cooking equipment, these visits also ascertain the sources of garbage fed and the general animal health status of livestock. An inspection cost estimate of $1200 per farm per year is used in this study assuming that the federal inspection cost estimate is based on APHIS's 1995 cost outlay on the Swine Health Protection Program (USDA 1995d)[19] and the state-administered program costs are similar to those in Florida.[20]

Assessment of Benefits/Costs of PFW Feeding at Current Levels

The set of assumptions concerning the technical and behavioral relationships in this industry are now applied to construct a 1993 baseline of the economics of PFW-feeding to swine in New Jersey, Florida, and the United States. This baseline relies on one more additional important assumption—that the same number of hogs that were PFW-fed in 1993 would have been raised using conventional production practices. Although important, this assumption also could have been incorporated easily into the analysis by assuming a production effect due to feeding regimes. In the present framework, we simplify the analysis so that regardless of the type of feed used, the same amount of hogs (and associated manure and odor) and pork (and associated benefit to consumers) is produced. This discussion is broken down into three parts: (1) the benefits and costs derived from the society-at-large, (2) those benefits accrued and costs incurred by producers, and (3) total benefits and costs from an economy-wide perspective.

Society-at-Large Benefit and Cost Estimates

Solid Waste Disposal Savings. As was discussed, tipping charges vary over time and regionally depending on many factors. A $30 per ton and a $75 per ton tipping charge were used as a low and high charge for the United States.[21] Multiplying these landfill tipping charges by the volumes of PFW fed nationally (547,500 tons) produced estimated savings of $16.43 and $41.06 million respectively (table 7.2). Florida and New Jersey serve as vivid examples of the range found in high-cost areas. For example, in Florida, rates vary from $23.50 per ton in Leon County in

Northern Florida to $125 per ton in Key West (Lawrence et al. 1988). Assuming the most interest in PFW-feeding is in areas charging higher tipping charges, this analysis used two relatively high tipping charges, $60 and $120 per ton for Florida. Based on the PFW-usage data (17,413 tons), multiplied by these two charges, produced estimated savings from $1.04 million to $2.10 million per year. New Jersey, one of our most populous states, is increasingly putting severe pressure on its land and other natural resources. This is evident in escalating tipping fees, which have increased from $15 per ton in the early 1980s to more than $125 per ton in the early 1990s, and its need to export more than 2 million tons of its municipal solid waste out of state each year (Personal Communication with Mr. Kevin Sullivan, 1995; Derr 1991b). For purposes of this analysis, two tipping fees ($50 and $100 per ton) used and multiplied by the estimated 20,800 tons of PFW fed to swine in New Jersey, gave $3.29 million and $6.58 million in solid waste disposal savings.

Inspection Costs. Inspection costs are the only direct cost incurred by society in permitting PFW-feeding operations. This per farm cost has been estimated to be $1,200. Multiplying this estimate by the number of licensed garbage-feeders in the United States (5,303), Florida (309), and New Jersey (31) produces the following estimates of aggregate social cost of inspection for each region—$6.36 million for the United States, of which $370,800 was spent in Florida and $37,200 in New Jersey.

Weighing Society-at-Large Benefits and Costs. The benefit/cost framework indicates huge social returns resulting from each dollar of social cost (inspections). On the social ledger, the savings in avoided landfill costs simply swamp inspection costs. This is easily seen in the extreme case of New Jersey with very low farm numbers (and subsequently low overall inspection costs). Each dollar of inspection costs yields $164 in social benefits based at $100 tipping charges (table 7.2). This estimate drops to $82 per $1 of inspection costs with a $50 per ton tipping charge. For Florida and the nation as a whole, the return for each dollar of inspection cost is still substantial—$2.81 to $5.65 for Florida and $2.58 to $6.46 for the United States (depending on whether the low or high tipping charge is used). These lower estimates are probably more indicative of what could be expected with increased PFW-feeding. However, if the trend toward fewer hog operations, as evident in national statistics, holds for PFW feeders, increased PFW could occur on fewer operations and not result in any increase in social costs.

Table 7.2. Aggregate costs/benefits of PFW-feeding to swine at current levels: New Jersey, Florida, and the U.S. (millions of dollars)

	Region/State					
	New Jersey Tipping fee of:		Florida Tipping fee of:		U.S. Tipping fee of:	
Item	$50/ton	$100/ton	$60/ton	$120/ton	$30/ton	$75/ton
A. "Society-at-Large" Benefit and Cost Estimates (millions of dollars)						
1. Solid waste disposal savings	3.29	6.58	1.04	2.09	16.43	41.06
2. Inspection costs	0.04*		0.37*		6.36*	
Net benefits to society	3.25	6.54	0.67	1.72	10.06	34.70
B/C ratio (Benefits per $1 Cost)	82.25	164.50	2.81	5.65	2.58	6.46
B. Producer-Specific Benefit and Cost Estimates (million of dollars)						
3. Producer feed cost savings	2.21*		0.90*		15.06*	
4. Net handling costs of PFW	0.16*		0.04*		1.37*	
5. PFW-feeding inefficiencies	0.29*		0.08*		1.86*	
6. Price discounts	1.09*		0.45*		7.43*	
7. Production effect	0.15*		0.06*		1.00*	
8. Sanitation measures	0.00*		0.03*		0.53*	
Total producer "costs"	1.69*		0.66*		12.19*	
Net benefits to producers	0.52*		0.24*		2.87*	
B/C ratio (benefits per $1 cost)	1.31*		1.36*		1.24*	
All benefits (1 plus 3)	5.50	6.58	1.94	2.09	31.49	41.06
All costs (2,4,5,6,7, and 8)	1.73*		1.03*		18.55*	
Net benefits	3.77	6.58	0.91	2.09	12.94	41.06
Overall B/C ratio	3.18	5.08	1.88	2.90	1.70	3.02

* Items not affected by changes in the level of tipping fees.

"Producer-Specific" Benefit and Cost Estimates

Net Producer Feed Cost Savings. The main benefit to producers in feeding PFW is the avoidance of using purchased commercial feeds. For the United States, if the 300,000 PFW-fed hogs produced in 1993 were fed conventional hog rations, an additional 4.3 million bushels of corn and 22,500 tons of soybean meal would have been required. Based on average 1993 prices, this amount of corn and soybeans was worth $15.05 million ($10.71 million of corn and $4.34 million of soybean meal) (USDA, 1995a).[22] Although this represents a small volume compared with the total markets for corn and soybean meal, less than one-half of 1%, this amount of crop production represents a harvest from nearly 42,600 acres of corn and 31,400 acres of soybeans. Of this total savings, $0.9 million in lower feed costs was realized in Florida (in lower corn use of 257,143 bushels and 1,350 tons of soybean meal). In New Jersey, a feed savings of 628,571 bushels of corn (valued at $1.6 million) and 3,300 tons of soybeans (valued at $0.6 million) was estimated.

Additional Handling Costs. As previously discussed, the only cost that could be allocated to PFW-feeders is the food-waste cooking costs. This assumes that the $5 per ton heating cost applies to 50% of all garbage fed in 1993 and adds $1.37 million in total operating costs.[23] Similar logic was applied to the situations in Florida and New Jersey.

PFW Feeding Inefficiencies. Cost budgets suggest a monthly building rental charge of $3.10 per hog per month. For the United States, multiplying the two-month overusage times $3.10 by 300,000 hogs gives a $1.86 million imputed cost assessed to hogs being fed PFW, due to their extended time in hog buildings. This same method was applied to Florida and New Jersey. In Florida, a monthly cost estimate of $2.27 per hog for the southern United States is assumed to be representative of costs times the numbers of PFW-fed hogs, which gives an additional $81,720. For New Jersey, a monthly cost estimate of $3.27 for the northern United States applied to New Jersey's PFW-fed swine numbers, multiplied by 2 months gave an estimated $287,760 cost.

Price Discounts. As previously discussed, an $8.26 per hundred price discount was applied to PFW-fed hogs in this study. For the United States, this discount multiplied by an average market weight of 300 pounds, multiplied by the estimated 300,000 PFW-fed hogs, produced a $7.4 million price penalty. Applying the same price discount on Florida's and New Jersey's PFW-fed hogs produced estimated losses of $446,040 and $1.09 million, respectively.

Disease Risk. The disease risk-related producer costs include any detrimental production effects brought about by feeding PFW and any costs incurred to safely use PFW. Recall that inspection costs are assumed to fall on society. The assumed production effect generated by a higher mortality rate for PFW-feeding operations (as opposed to grain-based operations) generated an annual loss of 9,000 hogs on PFW-feeding operations. This estimate multiplied by the 1993 average hog price ($37.06 per cwt.) gives an additional cost of increased mortality due to PFW-feeding of $1 million for the United States. Of the 9,000 hogs, an additional 540 and 1,320 hogs would have been lost in Florida and New Jersey, respectively, generating monetary losses of $60,000 and $146,760. Also, the modest cost outlay of $100 per year was applied to each operation in the United States and the two states to account for assumed additional sanitation measures taken by PFW-feeders. These costs totaled $0.53 million for the United States, $30,900 for Florida, and $3,100 for New Jersey.

Weighing Producers' Benefits and Costs. In the aggregate, PFW-feeding brought about positive returns to producers—$2.87 million in the United States, $520,000 in New Jersey, and $240,000 in Florida. While their feed cost savings are substantial, their implicit costs, not their out-of-pocket costs, were substantial. These producer-incurred costs consisted mostly of the implicit losses incurred by producers through feeding inefficiencies and price discounts. These economic costs, plus other reasons mentioned in the first part of this chapter, explain the low level of observed PFW-feeding. Still, for each dollar of producer costs, PFW-feeding hog producers received net benefits ranging from $0.24 to $0.36 as a result of lower feed cost outlays. In all cases, producer returns were sufficient to cover inspection costs if such costs were placed on a user fee basis and were incurred by producers, not society-at-large.

Economy-wide Benefit and Cost Estimates

Regardless of the landfill tipping charge assumed, "economy-wide" benefits exceed "economy-wide" costs. For the United States and the entire economy, a 1.70 BC ratio (assuming a $30 per ton tipping charge) was obtained. This means that for every dollar in costs throughout the system, $1.70 in benefits was realized through PFW-feeding (bottom row of table 7.2). With a $75 per ton tipping charge, the BC ratio increased to 3.02.

The relative distribution of benefits and costs from PFW-feeding across producers and the society-at-large poses an interesting question: how to encourage appropriate private actions that led to an apparent, sizable common good derived from PFW-feeding. In the current U.S. PFW-feeding situation, regardless of the landfill tipping charge, about one-third of total costs is incurred by society ($6.36 million in APHIS and state inspection costs) while about two-thirds ($12.2 million) is incurred by PFW-feeders. Higher tipping fees leave producer costs and returns unchanged and only increase the benefits accruing to society, causing the relative share of benefits from PFW-feeding to be skewed away from producers. When tipping fees are assumed to be $30 per ton, about 78% of the net benefits ($10.1 million of $12.9 million) accrue to the society-at-large and about 22% ($2.9 million) are realized by producers. However with a $75 per ton tipping charge, society-at-large net benefits balloon to $37.6 million, increasing its share of net benefits to 92%. Because producers' net benefits stay at $2.9 million, its relative share of net benefits shrinks to 8%. This kind of skewing may create a situation where society has a role in fostering some kind of market innovation (or public-funded mechanism) to facilitate improved PFW-processing and delivery systems. For example, New Jersey authorities have explored the economics of a "central corridor system" to facilitate the movement of PFW from the populous northern part of the state to the garbage-feeding hog operations in the southern part of the state. Results of this study indicated that collection costs associated with such programs were substantial and approached the value of the tipping charges avoided (Price et al. 1985).

Another approach could focus on creating a viable market for food waste by forcing food-waste generators to look seriously at other outlets. One such step could be to simply stipulate in law that all food waste from institutions that generate more than a specific amount of food waste per week be recycled, such as in Denmark, whose weekly threshold amount is 100 kg, (Skajaa 1989). This may help jump start a viable PFW market and help allocate it to swine producers if they can bid it away from other potential outlets. It must be kept in mind that PFW-feeders would be very sensitive to any higher costs associated with PFW. In any case, government regulations should not deter the establishment of a PFW market or technology that makes safe, cost-effective PFW products. There is some indication that current regulations may inadvertently affect technological innovations in PFW-processing by forcing cooking on relatively safe and precooked homogenous food waste.

Conclusions/Implications

Economic Viability. Under current conditions, PFW- feeding to swine occurs on the margin of profitability. Conventional feeds are relatively cheap, traditional production practices are the norm, and PFW-feeding occurs only in a few niche markets. PFW-feeding appears profitable for those producers that can manage the special handling and sanitation aspects of such production systems. Increased landfill tipping charges do not directly affect producers' decisions to feed PFW. The main producer decisions involve producers' attitudes concerning PFW, its availability, relative feed costs, and a host of other reasons outlined in table 7.1. Greater use of PFW is envisioned given new technology that could produce a dehydrated PFW product with increased storage capacity, more stable nutritional quality, and greater health assurances. Such a product could eliminate much of the herd health safety concerns that are behind many current regulations. These new technologies appear to be the key for increased use of PFW in the production of high-quality feed products for swine producers and diversion of potentially large amounts of food waste from landfills.

Social Benefit Creating Activity. The current situation can be viewed as the opposite of pollution problems created by private industries. In those cases, private actions do not include all their costs of production (that is, they exclude the clean-up charge and discomfort costs of pollution). Action by the government normally involves steps to force these firms to internalize all costs of these spill-over effects, e.g., prescribed or recommended production practices, taxes based on pollution created, subsidies on pollution abatement measures, etc. In the present case, it is the benefits, not the costs, to society that are not included by the economic agents involved. Therefore, some type of government action may be necessary to realize greater social welfare gains.

Disease Risk Potential. The current federal regulations establish one common set of measures to handle animal and human health safety concerns. This "blanket" regulation may have been necessary in the past and still relevant in certain cases where the origin of garbage used to make the PFW product is not known. However, in cases where food type and sources are known to be safe or where a processing technology (other than specified in regulations) can render it safe, some variant of current regulations may be justified.

Regulatory Dilemma. Current federal regulations fail to recognize the diversity of garbage sources and types and their potential to introduce and/or spread diseases, as well as the diversity of garbage-feeders. Even with increasingly concentrated sources of relatively safe food waste and rising landfill costs, hog feeders may not see any economic justification to feed PFW if current feed costs are lower than comparable costs of PFW. Some form of government intervention may be called for to divert increased amounts of food waste from landfills and encourage safe use of it as a swine feed. Two steps were discussed in this chapter: a direct facilitating role in centrally handling food waste and another that would mandate food-waste recycling. Current regulations probably have been beneficial in establishing industry standards for safe handling and feeding of PFW to swine by providing minimum processing guidelines to ensure the production of safe and healthy PFW-fed pork. However, present food technology may now be able to render safer PFW products than in the past. Also, the structure of the industry has become more concentrated, and it may make the handling and processing of PFW more feasible. Such changes may mean that current regulations do not apply as universally throughout the industry as in the past.

References

Acor, Geneva. 1994. Garbage Feeding Not All Waste, Written comments by Dr. Geneva Acor, University of Florida.

Derr, Don A. 1991a. Expanding New Jersey's Swine Industry Through Food Waste Recovery. Getting The Most From Our Materials: Making New Jersey The State-of-the-Art. June 1991.

Derr, Don A. 1991b. Economics of Food Waste Recycling. A Presentation to the Conference of Recycling Economics and Marketing Strategies. 1991.

Federal Code of Regulations (CFR), Public Law 96-468-October 17, 1980. Also referred to as the "Swine Health Protection Act."

Lawrence, J. D., M. L. Hayenga, and M. H. Jurgens. 1988. Feed Utilization Estimates for Livestock and Poultry in the United States. Miscellaneous publication. Iowa State University. Ames, IA.

Northeast-Midwest Economic Review. 1995. The State of Garbage: The Northeast-Midwest Leads the Nation in Recycling. Vol. 8, No. 6. July 1995.

Personal communication with Mr. Sullivan, Foreman in the New Jersey Department of Solid Waste Planning Department. February 1995.

Prest, A. R. and R. Turvey. 1965. Cost-Benefit Analysis: A Survey, *The Economic Journal*, Vol. 75, No. 300:683-705.

Price, A. T., D. A. Derr, J. L Suhr, and A. J. Higgins. 1985. Food Waste Recycling through Swine. *BioCycle Magazine* Vol. 26, No.2, pg. 34-37. March 1985.

Purdue University. 1976. Pork Production Systems with Business Analysis, ID-123, Cooperative Extension Service, West Lafayette, Indiana.

Purdue University. 1975. Troubleshooting the Swine Operation: A Guide for Evaluating Your Production Management Practices, AS-420, Cooperative Extension Service. West Lafayette, Indiana.

Shapouri, H., K. H. Mathews, Jr., and P. Bailey. 1994. Costs and Structure of U.S. Hog Production, 1988-91. USDA. Economic Research Service. Agriculture Information Bulletin Number 692.

Skajaa, J. 1989. Food Waste Recycling in Denmark. *Biocycle: J. Waste Recycling*. 30(11):70.

U.S. Department of Commerce. 1996. *Annual Survey of Manufacturers*. Bureau of the Census,, M96 (AS) -2.

U.S. Department of Commerce. 1992. Agricultural Census of the United States. Economics and Statistics Administration, Bureau of the Census. October 1994.

USDA. Internal APHIS national survey of statewide programs related to PFW feeders.

USDA (U.S. Department of Agriculture). 1995-1999. Agricultural Statistics. National Agricultural Statistics Service.

USDA (U.S. Department of Agriculture). 1995-1999. Hogs and Pigs Report. National Agricultural Statistics Service.

USDA (U.S. Department of Agriculture). 1995-1999. Meat Animals: Production, Disposition and Income Summary. National Agricultural Statistics Service.

USDA (U.S. Department of Agriculture). 1995a *Agricultural Outlook.* September 1995.

USDA (U.S. Department of Agriculture). 1995b. Economic Indicators of the Farm Sector, Cost of Production, 1993—Major Field Crops and Livestock and Dairy. Economic Research Service, ECIFS 13-3, July 1995, Table 14A, pg. 44.

USDA (U.S. Department of Agriculture). 1995d. Veterinary Science Division working paper on estimated APHIS Swine Health Protection Program. Animal and Plant Health Inspection Service.

USDA (U.S. Department of Agriculture). 1994. Agricultural Statistics, 1994.

USDA-APHIS. (U.S. Department of Agriculture). 1995c. Risk Assessment of the Practice of Feeding Recycled Commodities to Domesticated Swine in the U.S. Animal and Plant Health Inspection Service, Veterinary Services. Centers for Epidemiology and Animal Health. Fort Collins, CO.

Van Loon, Dirk. 1988. Small-Scale Pig Raising. Pownal, Vermont: Storey Communications, Inc.

NOTES

1. Areas in the U.S. that meet all or some of these conditions are Las Vegas, NV; Atlantic City, NJ; and many areas in Texas, Florida, and Puerto Rico.

2. Net benefits to food waste feeders are realized through lower feed costs — and possibly some revenue from fees received in removing food waste from food-waste generators — minus the direct costs of food waste pick-up and handling and several indirect costs as described in this chapter.

3. Items in parentheses are cited in the References section at end of this chapter.

4. Current federal regulations define garbage as food-waste items containing meat products and require it to be boiled for one-half hour before feeding it. This definition exempts food-waste products generally free from meat products, such as bakery and candy waste, from the cooking requirement on the basis that they pose little disease risk. Main oversight mechanisms are licensing of garbage-feeding operations, setting minimal standards for cooking apparatus and cooking guidelines, and periodic inspections of licensed premises to ensure that apparatus meets standards and to check the health status of the herd.

5. Federal regulations provide minimal guidelines in this area. Local and state governments have the authority to be even more restrictive.

6. The most recent and complete data on PFW-feeding to swine are limited to the early 1990s.

7. Finishing here refers to the 4- to 5-month process of "feeding out" young pigs weighing 40 to 50 pounds to slaughter weights of 220 to 235 pounds

8. By early 1999, the number of hog operators had dropped by 45% — to 114,840 — while maintaining or, at times, having higher total U.S. hog production.

9. This study assumes swine feeding is the most viable outlet for food waste. Similar studies on other alternative outlets — such as "waste-to-steam" energy recovery, composting, feeding to other livestock species — may be as feasible and beneficial to society, but no economic data could be found in the literature.

10. At first blush, it would appear that PFW-feeders could charge collection charges *plus* the previous tipping charges that are now "saved." Competition dissipates any possibility of these "phantom charges." Likewise, any lower food-waste disposal fees on public institutions, such as public schools and prisons, may be realized as simply *illusionary*. These institutions' lower fees are now based on only collection and handling charges and are appropriately not considered part of the value of the social benefits from saved landfill space.

11. Tipping charges are fees levied on truckers to discharge their load ("tip" their dump truck) at landfills.

12. Sensitivity analysis on the data used in this study found that as the level of nonedibles in the food waste increases, the profitability of food waste in swine feeding decreases. For example, nonedible contents exceeding 15% renders food-waste feeding uneconomic at $30 per ton tipping fees. At higher tipping fees, the threshold nonedible content level is reduced to as low as 6.5%.

13. Any costs specific in processing garbage into a suitable PFW swine feed and that would not be needed if conventional feeds were being fed, would need to be deducted from any savings. This concern is considered in the first section under "costs."

14. This study assumes a feeding rate of almost one ton of PFW per hog or almost 10 pounds of PFW per one pound of gain.

15. This through-put is the level recommended for high-capital intensive operations in the Midwest (Purdue University 1976).

16. The average 1993 slaughter price was $45.32 per cwt for barrows and gilts and $37.06 per cwt for sows as reported on table 405, pg. 241.

17. Some PFW-feeders may feed out their hogs to these heavier weights for a number of reasons, including low POC and FOC, feed-to-meat conversion on their hogs that may not markedly decrease with age

and body weight, a price penalty that may exist on PFW-fed hogs regardless of their weight, and the preference of older hogs that have been proven durable survivors and can be depended upon to process large amounts of food waste.

18. If such an outbreak occurred, more far-reaching, macro effects would need to be considered in this study, such as impacts on future production, on trade, and on disease containment and/or eradication program costs.

19. APHIS's costs divided by the number of licenced garbage feeders (5,303) gives an estimated $600 per operation. Thus, this estimate may overstate true costs, as it also includes costs of other programs and searches of other garbage-feeding farms that may be operating without a license.

20. A 1994 presentation on garbage-feeding in Florida reported that Florida's inspection program (involving several farm visits per month) incurred $350,000 in costs (Acor 1994). The author stated that the Florida Animal Industry Bureau estimated that program costs could be reduced to $200,000 if inspection visits moved to a single monthly visit. Dividing the current and reduced program cost estimates by the number of Florida garbage feeders (309) gives a high and low range for per farm inspection charges: $1,133 to $647.

21. Although the U.S. average landfill tipping charge was $30 per ton in 1993 (Northeast-Midwest Economic Review 1995), it is assumed that $30 per ton is more on the low end in areas where food-waste feeding occurs.

22. The average 1993 U.S. price for corn was $2.50 per bushel and $193 per ton for soybean meal as reported on table 17, pg. 43.

23. It is assumed that 50% of food waste fed consists of exempted food-waste products, such as bakery and candy waste, and that some portion of cooking is done by using lower cost fuels, for example salvage wood.

8

Dehydrated Restaurant Food Waste as Swine Feed

by R. O. Myer, J. H. Brendemuhl, and D. D. Johnson

On a dry basis, restaurant food wastes (plate wastes) are high in nutrients desirable for pig feeding. Typical analyses (dry matter basis) previously reported include crude protein contents of 15 to 23%, fat (ether extract) of 17 to 24%, and ash of 3 to 6% (Kornegay et al. 1970; Pond and Maner 1984; Ferris et al. 1995; Westendorf et al. 1996). Traditionally, restaurant food wastes have been fed to pigs with little processing. During the past few decades, however, the feeding of wet food waste (or garbage-feeding) has decreased. This decrease has been attributed to increased regulation, the outright ban of feeding food wastes to pigs in many states, the high labor requirements involved, relatively poor performance of food waste-fed pigs, the movement to large consolidated swine operations, the availability of low cost alternative waste disposal outlets (i.e., landfills), and the undesirable stigma associated with food-waste feeding operations (Westendorf 1996; Westendorf et al. 1996).

In many areas of the United States, however, waste disposal options, such as landfills, for food waste are becoming more expensive and scarce. Thus, recycling food waste for livestock feeding, once again, is a viable waste disposal option. This high disposal cost, along with advancements in dehydration technology, has prompted interest in the dehydration of restaurant food waste. The advantages of dehydration are obvious, and furthermore, the dehydrated end products could easily be incorporated into many pig-feeding programs already in existence today.

The term "restaurant food waste" or simply "food waste" will be used to indicate waste from restaurants, hotels, and other food preparation and food service establishments, and to differentiate from food waste associated with food-processing industries (i.e., meat packing, citrus juice, vegetable canning, etc.).

113

The dehydration of much food waste from food processing is rather common and has been occurring for quite some time. For example, in Florida, commercial production of dried citrus pulp began in the early 1930s. Citrus pulp is the main by-product feedstuff from the citrus processing industry and has been shown to be an excellent feedstuff for both dairy and beef cattle (Ammerman and Henry 1991). Until recently, dehydration of restaurant food waste has not been an economical and viable alternative. This has been due mainly to its high moisture content (60 to 90%) and heterogenous nature, lack of very large generators, and the availability of inexpensive waste disposal options. Even with advancements in drying technology, the dehydration of restaurant food waste is and would be rather expensive. However, if the potential dehydrator can recoup a portion of the tipping fee or landfill disposal fee to subsidize the costs involved, then dehydration may be economical, especially in those areas with high landfill disposal fees.

Basic Dehydration Process

Dehydration procedures involved in recent research have used two basic strategies/processes. Other processes may have been used or are under development of which the authors are unaware. Other dehydration methods, such as direct fire drying, have been used but generally for homogenous food waste such as bakery waste.

The two dehydration procedures involve the blending of minced food waste with a dry feedstock before drying. The first method involves simple dry extrusion. In this process, the minced food waste is blended with a dry feedstock such that the resulting blend is about 25% moisture. The semi-moist blend is then forced through an extruder. Heat is generated by pressure and friction upon forcing the mixture through the extruder. After exiting the extruder, the heated product cools and moisture is lost. The resultant "cooked" product contains about 13 to 16% moisture. The final product, however, is mostly feedstock and typically, on a dry basis, contains just 10 to 15% food waste. The extruder, however, can be coupled with a fluidized bed dryer. This would allow the extrusion of slightly higher moisture blends (i.e., 30% instead of 25%), thus resulting in a higher concentration of food waste in the finished product (i.e., 15 to 25%).

The second process involves a low temperature dry extruder/pelleter coupled with a high heat, high airflow, fluidized bed dryer. In this process, the minced food waste is blended with a dry feedstock such that the resulting blend is about 40% moisture. The semi-moist blend is passed through the extruder/pelleter. The resulting semi-moist pellets are then passed through a dryer where the pellets literally dance across hot air. The pellets exit the dryer and are cooled. With this process, the final product contains about 25 to 30% food waste (dry basis). In both processes, the final product can be used as the dry feedstock for subsequent dehydration. This "recycling" would result in a higher concentration of food wastes in the final product. While this concentration would be higher, this would have to be balanced against higher drying costs and a higher chance of nutrient destruction brought about by the increased exposure to high temperatures. In the extrusion-alone process, temperatures of 120 to 150°C (250 to 300°F) are reached, and the process is rather instantaneous, lasting only 15 to 30 seconds. In the second, temperatures of 100 to 120°C (210 to 250°F) are typically reached in the product being dehydrated, and the process lasts about 2 to 7 minutes.

In both processes outlined above, a dry feedstock is used to aid in the dehydration process. Most any dry feed material can be used. Some feedstocks that have been successfully utilized include wheat middlings, finely ground corn, soy hulls, ground peanut hulls, soybean meal, and rice hulls. Previously dried food-waste products have also been utilized. Many factors will influence the type of feedstock to be used, including that the feedstock (1) should be readily available, (2) is in a form that does not require further processing prior to use (i.e., grinding), (3) produces an end product that can be handled easily and stored, (4) enhances feed value, (5) should be appropriate for the class of livestock to be fed (i.e., pigs v. dairy cattle), and above all is 6) economical to use. To date, the most widely used feedstock appears to be wheat middlings.

Both processes outlined above have been successfully utilized in the dehydration of other high-moisture materials into nutritious feedstuffs. Examples of raw products used include catfish-processing waste (Tacon and Jackson 1985), scallop viscera (Myer et al. 1987), poultry mortalities (Blake et al. 1991; Tadtiyanant et al. 1993; Myer 1998), "spent" hens (Haque et al. 1991; Lyons and Vandepopuliere 1996), and poultry hatchery waste (Tadtiyanant, et al. 1993).

Initial Research with Dehydration of Restaurant Food Waste

Researchers at the University of Florida have been involved in the evaluation of dehydrated restaurant food-waste products (DFW) since the early 1990s. A preliminary study, conducted in conjunction with Azurea Inc., a central Florida environmental engineering consulting firm, was done to assess a dehydrated food-waste product as a potential feedstuff for pigs (Myer et al. 1994; Rivas et al. 1995). This initial assessment included (1) determination of nutrient composition and laboratory quality assessments of important nutrients, (2) determination of digestibility by pigs, and (3) conduction of microbiological safety assessments.

For this pilot study, food waste was obtained from food service operations of two hotels at a resort complex in central Florida. The waste was mostly leftover food and plate scrapings. A total of 20 55-gallon drums was collected. The contents of the drums were blended and a subsample taken to fill three drums. The three drums were sealed and shipped to Jet Pro Inc. in Atchinson, Kansas. The contents of the drums were then dried using a Jet Pro dryer (the second process outlined above). Drying air temperature was maintained at 170 to 190°C (360 to 400°F) in which the product temperature reached 95 to 115°C (200 to 240°F). Transit time through the fluidized bed dryer was about 3 minutes. A small amount of the food waste was initially dried after blending with a dry feedstock (soybean meal). This dried product was the feedstock for subsequent drying. This process was repeated several times. After drying, the product was shipped back to Florida.

The resulting DFW product was essentially pure dried food waste. This product was a pelleted, dark brown product that was greasy to the touch. The material had a mild odor that could be characterized as a combination of feed fat, fish meal, ground grain, and coffee grounds.

Initial Nutritional Composition Results

Composition of the major chemical components of this initial DFW product is given in table 8.1. For comparative purposes, corn and soybean meal were analyzed and results are also given in table 8.1. The analyses indicated the DFW to be quite dry, moderately high in protein, high in fat, and relatively low in crude fiber and ash (total mineral matter). The relatively high protein and fat contents and the low fiber and ash contents are desirable in a feedstuff for pigs. The chloride and sodium contents, however, were high, indicating a high salt content. These high-sodium and -chloride contents could limit the amount of this DFW product that could be included in pig diets to still achieve good hog growth performance. Salt is commonly added to pig diets, thus the high salt content of DFW could replace salt supplementation.

Since DFW was found to be moderately high in protein, further analysis of the protein was conducted and is summarized in table 8.2. In addition to quantity, the quality of the protein is very important in pig feeding as well as for other simple stomached animals (i.e., poultry, fish,

Table 8.1. Composition of major components of dehydrated restaurant food waste (DFW) and other representative feedstuffs[a]

Item	DFW	Corn	Soybean meal[b]
Moisture	7.9	10.80	11.60
Crude protein	22.4	9.30	48.10
Crude fat	23.2	3.70	0.80
Crude fiber	2.3	2.90	3.50
Total mineral matter[c]	5.4	1.30	6.60
Calcium	0.5	0.02	0.30
Phosphorus	0.5	0.30	0.70
Chloride-soluble	1.3	0.04	0.03
Potassium	0.7	0.40	2.20
Sodium	0.9	<0.005	0.02

[a]Values are expressed on an as is basis. Each value is an average of duplicate analyses.
[b]Commercially available 48% crude protein product.
[c]Ash.

dogs, etc.). With one possible exception, the profile of the essential amino acids in the protein of DFW (expressed as g/100 g of total protein) was similar to the profile in soybean meal. Soybean meal is considered to be a good, quality protein source. The exception was lysine, which was lower in DFW than in soybean meal on a g/100 g protein basis. Soybean meal is considered to be high in lysine. Lysine is important in pig feeding because it is usually the most limiting of the essential amino acids in typical pig diets. Because of the lower lysine relative to soybean meal, protein quality of the DFW product would be considered only fair.

Two in vitro laboratory tests were also done to further assess the quality of the protein. These were pepsin digestibility and available lysine determination. The results of these determinations are given in table 8.2. Pepsin digestibility is an assay used to predict the digestibility of the protein. Pepsin digestibility of the protein of DFW was 73% digestible, considered just fair. The value obtained for soybean meal was 91%, which is considered very good. Available lysine is an assay procedure used to estimate the portion of the total that can be utilized for growth and metabolism. Lysine in a peptide chain of a protein has an exposed amino group (epsilon amino group). This amino group can readily react, especially under high temperatures, with other compounds (i.e., reducing sugars, oxidized fats) that would render the lysine unavailable (indigestible). This reaction is referred to as the browning reaction

Table 8.2. Amino acid composition of the protein and in vitro protein quality assessments of dehydrated restaurant food waste (DFW) and commercial soybean meal (%)[a]

Item	DFW	Soybean meal[b]
Crude protein, total	23.1	48.4
Pepsin indigestible protein	3.6	2.7
Essential amino acids:[c]		
Lysine	0.78 (3.4)	2.88 (6.0)
Available lysine	0.57 (2.5)	2.62 (5.4)
Threonine	0.92 (4.0)	2.10 (4.3)
Tryptophan	0.18 (0.8)	0.50 (1.0)
Methionine	0.42 (1.8)	0.69 (1.4)
Isoleucine	0.88 (3.8)	2.09 (4.3)
Valine	1.04 (4.5)	2.25 (4.6)
Leucine	1.56 (6.8)	3.50 (7.2)
Histidine	0.70 (3.0)	1.30 (2.7)
Phenylalanine	0.86 (3.7)	2.26 (4.7)
Arginine	0.85 (3.7)	3.20 (6.6)
Nonessential amino acids:[c]		
Aspartic acid	1.85	5.70
Serine	1.06	2.70
Glutamic acid	3.80	8.00
Proline	1.22	2.10
Glycine	1.01	2.05
Alanine	1.07	2.25
Cystine	0.26 (1.1)	0.64 (1.3)
Tyrosine	0.44	1.40

[a]Values are expressed on an as is basis (air dry basis). Each value is an average of duplicate analyses.
[b]48% soybean meal.
[c]Numbers in parentheses are amino acid levels expressed as g per 100 g of protein.

or sometimes called the "Maillard" reaction (Hurrell 1990). The amount of available lysine in DFW was estimated to be 85% of the total, which would be considered good, however, the estimate for soybean meal was 95%.

Since DFW was found to be high in fat, quality evaluations of the fat were done. Fat is a concentrated source of energy and a high level in a feedstuff would increase its value for use in pig feeding. The composition of the fat in DFW and results of quality assessments of this fat are shown in table 8.3. For comparison, a sample of a commercially available livestock feed fat was also analyzed and results presented in table 8.3. The fatty acid profile of the fat in DFW does not indicate any potential problems that might occur as a result of feeding this fat to pigs. The fatty acid profile was found to be actually more desirable than that obtained

Table 8.3. Composition and quality assessments of the fat in dehydrated restaurant food waste (DFW) and of a commercial livestock feed fat product[a]

Item	DFW	Feed fat[b]
Crude fat (ether extract) (%)	23.20	99.00
Fatty acid profile, relative (%)		
C6:0	0.50	<0.10
C8:0	0.40	<0.10
C10:0	1.00	<0.10
C12:0	1.50	<0.10
C14:0	4.50	1.20
C14:1	0.40	0.20
C15:0	0.50	0.20
C15:1	0.10	<0.10
C16:0	21.80	18.00
C16:1	1.80	1.90
C17:0	0.40	0.50
C18:0	9.80	11.10
C18:1	34.00	45.50
C18:2	17.80	17.50
C18:3	2.00	1.30
C18:4	0.20	0.20
C20:0	0.20	0.30
C20:1	0.40	0.50
C20:2	0.20	0.20
C20:4	0.40	0.10
C20:5	0.20	<0.10
C22:6	0.30	<0.10
Other	1.00	0.50
Total saturates	40.80	31.70
Total monounsaturates	37.10	48.40
Total polyunsaturates	21.10	19.30
Unsaturated:saturated ratio	1.42	2.14
Peroxide value-initial, meq/kg[c]	3.00	1.80
TBA[d] rancidity, mg/kg[e]	1.55	0.46

[a]Expressed on as fed basis; average of duplicate analyses.
[b]Mixture of tallow and reclaimed restaurant grease (brown grease).
[c]Units of peroxide formation per kg of fat.
[d]Thiobarbituric acid.
[e]Mg malonaldehyde per kg of fat.

for the feed fat. The DFW fat was more saturated overall. Thus, problems with resulting soft fat in the carcass fat of pigs fed this product should be minimal. Diets high in unsaturated fat when fed to pigs can result in carcasses with soft fat, which are undesirable because of difficulties with processing and merchandising these carcasses (West and Myer 1987). Although the fat in DFW was highly saturated, the peroxide assay, which measures the stability of the fat, indicated some stability (oxidation) problems. The values obtained, however, were rather modest and would be of little concern. The TBA rancidity assay, which measures rancidity development, also obtained modest values indicating only slight rancidity.

Initial Apparent Digestibility Results

Table 8.4 summarizes the apparent digestibility coefficients of dry matter, energy, and crude protein of experimental diets that were obtained in a digestibility trial with young, growing pigs (20 to 40 kg avg. body wt.). Nutritionally adequate diets containing 0, 5, 10, or 20% DFW were fed in this trial. The digestion coefficients for dry matter, energy, and crude protein were similar (P > 0.05) among diets containing DFW. However, the 10% DFW diet had lower (P < 0.05) digestible energy and dry matter coefficients when compared to the diet with no DFW. All three diets containing DFW had lower (P < 0.02) digestible crude protein coefficients compared to the diet with no added DFW. However, no further depression in protein digestibility was noted with increasing levels of DFW from 5 to 20%.

Other than the high salt content, the potential negative findings of this initial study (i.e., slightly lower pepsin digestibility, lysine availability and protein digestibility, and slight rancidity development) may be due to processing and/or storage conditions rather than the makeup of DFW. The dark color of the pelleted DFW indicated some charring occurred during drying. Overheating of feed materials is known to decrease their digestibility and can cause rancidity development in the fat portion (Hurrel 1990; Zhang and Parsons 1996).

Table 8.4. Apparent digestibility coefficients of diets containing dehydrated restaurant food waste (DFW) for growing pigs (%)[a]

	% DFW in diet[b]				
	0	5	10	20	SE[c]
Dig. dry matter	90.1[d]	87.5[d,e]	84.2[e]	87.4[d,e]	1.1
Dig. energy	90.4[d]	87.2[d,e]	83.8[e]	87.2[d,e]	1.1
Dig. crude protein	88.3[f]	81.6[g]	81.7[g]	81.9[g]	1.8

[a]Each value is a mean of data from four animals; approximate average pig weight, 15 to 30 kg.
[b]Corn/soybean meal-based diets with DFW added in place of corn and soybean meal to maintain similar levels of protein and energy across all diets (crude protein ~ 20% and gross energy ~ 4,700 kcal/kg).
[c]Standard error; n = 4.
[d,e]Means in the same row with different superscripts are different ($p < 0.05$).
[f,g]Means in the same row with different superscripts are different ($p < 0.02$).

Results of Initial Microbiological Testing

Microbial analysis of DFW samples from the above initial trial indicated only low numbers of bacteria and other microbes present. The low level was probably due to the heat treatment the DFW was exposed to during processing. Table 8.5 summarizes the effect of storage length on total microbial population on samples of DFW kept in sealed paper bags under environmental conditions similar to that of a feed mill. Over a period of 2 weeks, the microbial population increased as conditions became more favorable for microbial (bacterial) proliferation. Nonetheless, microbial populations remained minimal and did not exceed those usually encountered for other feedstuffs such as corn or soybean meal.

Data is limited on the microbiological safety of dehydrated restaurant food-waste products. Using extrusion, Walker and Kelly (1997) did not detect "indicator microbes" (total coliforms, fecal coliforms, and streptococci microbial "groups") above the minimum detection limit as long as temperatures exceeded 127°C (260°F) in tests involving dehydration of cafeteria food waste. With some test runs, they obtain good results with temperatures as low as 116°C (240°F). They concluded that while there was some evidence of postextrusion survival of microbial contaminants, the levels noted were substantially lower than that noted in production-run pig feed samples. Lyons and Vandepopuliere (1996) conducted dehydration trials with spent hens that utilized equipment similar to that of the above Florida research. In their research, ground spent hens mixed with wheat middlings (1:1, dry matter basis) was dehydrated at an air temperature of 180°C (360°F) with product temperature estimated to be 110°C (230°F). Ground raw hen was positive for *Salmonella,* but the dehydrated hen/wheat middlings product was negative for this organism as well as for coliform. While the data are preliminary, Pace (1997) noted acceptable microbial kills upon dehydration of restaurant food waste-blended products using Jet Pro equipment with product temperatures of 82°C (180°F) or above.

Table 8.5. Effects of storage time on the total microbial population of dehydrated restaurant food waste.

Time	Total plate count[a]
Week 0	2.93×10^2
Week 1	3.81×10^2
Week 2	5.47×10^2

[a]Average number of colony forming units/g of sample.

Total microbial counts from the feces of pigs fed the experimental diets during the above-mentioned digestion trial were determined and are reported in table 8.6. The total microbial count was actually higher from pigs fed the control diet (0% DFW) than for pigs fed the diets containing DFW. Table 8.7 summarizes the individual microbial populations in fecal samples from the pigs during the digestion trial. Organisms found were normal for the gut microflora of pigs. High populations of *Klebsiella, Escherichia coli,* and *Listeria* can lead to diarrhea in the pig. Furthermore, high counts of these organisms increase the chance of carcass contamination. *Yersinia enterocolitica,* a microorganism of concern to the swine industry, was present, but at a normal level. Thus, the feeding of the diets containing DFW appeared to have no particular effect on the microbial flora of the pigs.

Table 8.6. Total microbial counts in feces from pigs fed diets containing dehydrated restaurant food waste (DFW) during the digestion trial[a]

% DFW in diet	Total plate count[b]
0	4.9×108
5	1.3×106
10	2.9×107
20	8.7×107

[a]Fecal samples remained frozen for a period of two months prior to analysis.
[b]Average number of colony forming units/g of sample.

Table 8.7. Microbial flora in feces from pigs fed diets containing dehydrated restaurant food waste (DFW) during the digestion trial.[a]

	% DFW in diet			
Organism	0	5	10	20
Listeria	10^5	10^5	10^4	10^4
Enterobacter aerogenes	10^5	10^5	10^5	10^5
Yersinia enterocolitica	10^4	10^3	10^3	10^3
Staphylococcus aureus	10^4	10^4	10^3	10^4
Escherichia coli	10^5	10^5	10^5	10^5
Shigella	10^3	10^3	10^3	10^3
Salmonella	10^3	—	—	—
Streptococcus faecalis	10^5	10^6	10^4	10^5
Klebsiella	10^5	10^6	10^5	10^5
Molds and yeast	10^4	10^5	10^4	10^4

[a]Each column represents estimated number of colony forming units/g of sample.

Early Trials with Finishing Pigs

Two feeding trials involving finishing pigs (55 to 115 kg) were conducted by the University of Florida (Myer et al. 1998) on the evaluation
of dehydrated food waste products as potential feed ingredients in
mixed pig diets. These two trials utilized dehydrated products produced
during initial test runs at NutraFeed Inc. of Clermont, Florida. For each
trial, fresh food waste was obtained from food service operations at a
resort complex in central Florida. The food waste was mostly leftover
food and plate scrapings, and contained about 60 to 80% water. For
both trials, the food waste was minced and blended with a dry feedstock
(55:45 blend of soy hulls and surplus wheat flour for trial 1 (DFW1) and
67:33 blend of soy hulls and ground corn for trial 2 (DFW2)) such that
the resulting blend was about 40% moisture. The semi-moist blend was
then pelleted and dried. Drying temperature was 150 to 200°C (300 to
400°F) such that the product temperature reached 110 to 120°C (230 to
250°F); transit time through the fluidized bed dryer was 3 to 7 minutes.
Soy hulls were used as part of the feedstock because of their absorbency,
and it was felt that the pellets produced would easily dry. To increase the
concentration of food waste in the final dried product used for both trials, the initial dried product was blended with additional minced, fresh
food waste and dried. The final DFW products used in these trials contained about 60% dried food waste with the other 40% being the feedstock. The two batches (DFW1 and DFW2) of the DFW product were
produced at different times and each involved a different collection of
food waste. The waste was collected from the resort during the night,
and the dehydration took place the next day.

The first feeding trial utilized 48 crossbred pigs. Dietary treatments
consisted of corn/soybean meal-based diets containing 0% (control) or
40% DFW product (DFW1). The pigs were fed the experimental diets
from 63 to 112 kg average body weight. The second trial utilized 72
crossbred pigs that were fed the experimental diets from 56 to 108 kg.
Dietary treatments for the second trial also consisted of corn/soybean
meal diets containing 0% (control), 40%, or 80% DFW product
(DFW2). In both trials, the nutritionally adequate diets were formulated
following NRC (1988) guidelines and were similar in estimated
digestible lysine (estimated calorie to lysine basis) content within the finisher diet types. At the end of the feeding phase for both trials, all pigs
were slaughtered to obtain carcass composition and meat quality data,
including taste evaluations of broiled loin chops.

Each of the dehydrated restaurant food-waste products utilized in
these two trials was a pelleted, tan to dark brown-colored product that

was slightly greasy to the touch. The color was lighter than initially noted for the DFW used in the pilot study. The products had a mild odor that could be characterized as a combination of fish meal, feed fat, ground grain, french fries, and coffee grounds. The pellets were reground before mixing into the experimental diets.

Composition of the major chemical components of the two DFW products is given in table 8.8. For comparative purposes, soybean meal samples were analyzed and results are also given in table 8.8. Unlike the analyses of the preliminary study, the composition of the DFW products reflected dilution by the feedstocks used. Nevertheless, the analyses indicated the DFW products to be low in moisture, high in fat, and moderately high in protein, crude fiber, and ash. The chloride and sodium contents, however, were high. These high contents equate to a salt content of about 1 to 2% in the DFW products. In the finishing pig trials,

Table 8.8. Composition and in vitro quality assessments of dehydrated restaurant food waste products (DFW) and of other feedstuffs used in pig finishing trials 1 and 2[a]

Item	DFW1	DFW2	Corn	Soybean meal[b]	Feed fat[c]
Moisture (%)	11.40	8.40	11.20	12.00	0.20
Crude protein (%)	15.00	14.40	8.90	48.10	—
Lysine (%)	0.63	0.64	0.26	2.92	—
Available lysine (%)	0.56	0.53	—[g]	2.78	—
Threonine (%)	0.56	0.60	0.30	2.16	—
Isoleucine (%)	0.56	0.56	0.24	2.06	—
Pepsin indigestible protein (%)	2.20	3.20	—	3.60	—
Crude fat (%)	13.80	16.00	3.50	1.20	99.00
Total saturates (%)	37.00	35.60	—	—	35.10
Total monounsaturates (%)	38.00	40.10	—	—	48.20
Total polyunsaturates (%)	23.50	22.90	—	—	15.10
Peroxide value-initial (meq/kg)[d]	18.00	5.00	—	—	2.60
TBA[e] rancidity (mg/kg)[f]	6.00	1.60	—	—	<0.25
Crude fiber (%)	10.30	14.50	2.10	3.50	—
Total mineral matter (%)	5.80	4.70	1.10	6.40	—
Calcium (%)	0.54	0.63	0.02	0.30	—
Phosphorus (%)	0.34	0.38	0.26	0.72	—
Chloride-soluble (%)	0.69	0.86	0.05	<0.02	—
Potassium (%)	0.55	0.80	0.32	2.20	—
Sodium (%)	0.35	0.47	<0.01	<0.01	—

[a]Values, other than for moisture, were adjusted to 88% dry matter basis, except for feed fat which is on an as is basis.
[b]Commercially available 48% product. Average for both trials.
[c]Commercial livestock supplemental fat (brown grease). Average for both trials.
[d]Units of peroxide formation per kg of fat.
[e]Thiobarbituric acid.
[f]Mg malonaldehyde per kg of fat.
[g]Not determined.

salt was not added to the DFW diets; the estimated salt (NaCl) content of the DFW diets was 0.4 to 0.8% v. 0.3% for the control diets. The relatively high content of crude fiber was probably from the soy hulls used in the initial blendings with fresh minced food waste before dehydration.

Estimated composition of restaurant food waste, minus the feed-stocks, in the DFW products (dry matter basis) would be 24 to 26% crude fat, 18 to 20% crude protein, 4 to 7% crude fiber, 5 to 6% ash, about 0.6% calcium and 0.4% phosphorus, and 2.0 to 2.5% salt. These levels, including the high salt content, agree with that previously obtained with DFW in the preliminary study mentioned above and by others analyzing fresh and cooked food waste (dry matter basis) (Korne-gay et al. 1970; Pond and Maner 1984; Walker and Wertz 1994; Ferris et al. 1995; Westendorf et al. 1996).

The profile of the essential amino acids in the protein of the DFW products was similar to the profile in soybean meal (expressed in g/100 g of total protein basis) with the possible exception of lysine (table 8.8). Lysine concentrations were lower in DFW products than in soybean meal, but were still higher than in corn (NRC 1988).

Pepsin digestibility and available lysine in the DFW products were found to be moderately high, but lower than obtained for soybean meal (pepsin digestibility of 86 and 80% for DFW1 and DFW2, respectively, v. 92 and 93% for the soybean meals, and lysine availability of 89 and 83% v. 94 and 94%). The lower values were probably the result of the heating that occurred during dehydration, combined with the availability of reactive substrates (i.e., reducing sugars), resulting in some nonenzy-matic browning (Maillard reaction; Hurrell 1990).

The fatty acid profile of the fat in the DFW products and results of quality assessment of this fat are shown in table 8.8. The ratio of total unsaturated to saturated fatty acids in the fat of the DFW products was similar to that found in the commercial livestock feed fat. However, the percentages of polyunsaturated fatty acids were higher in DFW products. The fatty acid profile of the DFW products indicates a mixture of fats of animal and vegetable origin. The commercial feed fat product analyzed was a blend of animal and vegetable fats, primarily beef tallow and reclaimed cooking oils from restaurants. Unlike the profile noted in the preliminary study, the profiles showed a higher proportion of polyunsaturated fatty acids. This change is probably a reflection of the general shift over the last several years to greater use of vegetable oils in restaurant cooking.

Peroxide and thiobarbituric acid (TBA) numbers obtained for the DFW products indicate some rancidity development (table 8.8). By con-trast, the peroxide and TBA values obtained for the commercial feed fat were quite low and indicated very little rancidity development. The feed fat product, however, was a product stabilized by the addition of an antioxidant. The DFW products used were not stabilized.

The chemical analyses of the two DFW products used in these two trials were in general agreement (table 8.8). One of the problems with feeding food waste is the variation in types and sources that results in variation in composition (Pond and Maner 1984). The food waste used in each trial was taken at different times but from the same sources and processed similarly.

Growth rates obtained in both finishing trials were excellent. The average daily rate of weight gain of the pigs was not affected (P > 0.10) by the inclusion of the DFW1 product at 40% of the diet in trial 1 (table 8.9), or the DFW2 product at 40 or 80% in trial 2 (table 8.10). Pigs fed

Table 8.9. Performance of finishing pigs fed diets containing a dehydrated restaurant food waste product (DFW) trial 1[a]

	Dietary treatment		
Item	0% DFW1[b]	40% DFW1	SE[c]
Avg. daily gain (kg)	1.01	1.01	0.012
Avg. daily feed intake[d] (kg)	3.38	3.00	0.084
Feed/unit gain[e] (kg/kg)	3.35	2.98	0.087

[a]Four pens per treatment with six pigs per pen. On experiment from 63 to 112 kg average body weight per pig.
[b]Dehydrated restaurant food waste blended product (blended prior to dehydration); approximately 60% food waste (dry) and 40% soy hulls/wheat flour blend (55:45); DFW1.
[c]Standard error of the mean; n = 4.
[d]Means differ ($p < 0.05$).
[e]Means differ ($p = 0.06$).

Table 8.10. Performance of finishing pigs fed diets containing a dehydrated restaurant food waste product (DFW) trial 2.[a]

	Dietary treatment (% DFW2[b])			
Item	0	40	80	SE[c]
Avg. daily gain (kg)	0.91	0.91	0.90	0.016
Avg. daily feed intake[d] (kg)	2.98	2.88	2.67	0.036
Feed/unit gain[d]	3.27	3.17	2.98	0.025

[a]Three pens per treatment with eight pigs per pen. On experiment from 56 to 108 kg average body weight per pig.
[b]Dehydrated restaurant food waste blended product (blended prior to dehydration); approximately 60% food wastes (dry) and 40% soy hulls/ground corn blend (67:33); DFW2.
[c]Standard error of the mean; n = 3.
[d]Means differ ($p < 0.01$; linear).

the DFW diets in trial 1 required, on average, 11% less feed per unit of weight gain (P = 0.06) than pigs fed the control. The better feed-to-gain was likely due to the higher fat content of the DFW1 diets (7 v. 3%). Likewise, feed-to-gain improved in trial 2 upon inclusion of the DFW2 product in the diets (P < 0.01; linear).

Average backfat thickness and average loin eye area were not detrimentally affected (P > 0.10) by the inclusion of the DFW product in the finishing diets in either trial (tables 8.11 and 8.12). Likewise, average estimated carcass lean content was also unaffected (P > 0.10) by the inclusion of the DFW product in either trial. Lean color, firmness, and marbling scores were similar (P > 0.10) between the DFW- and control-fed pigs in either trial. Carcass fat from pigs in trial 2 became softer as the level of DFW2 increased in the diets (P < 0.01; linear). This softening effect is likely the result of the relatively high level of polyunsaturated fatty acids in the fat of the DFW2. This softening was minimal from pigs that were fed the lower level of DFW2 (40% of the diet).

The inclusion of the DFW products in the diets of finishing pigs from either trial had no effect (P > 0.10) on palatability of characteristics of broiled loin chops as determined by a trained sensory panel (tables 8.13 and 8.14). The values obtained indicated that the chops were acceptable in tenderness and juiciness and had no detectible off-flavor. The

Table 8.11. Carcass characteristics of finishing pigs fed diets containing a dehydrated restaurant food waste product (DFW) trial 1[a]

	Dietary treatment		
Item[b]	0% DFW1[c]	40% DFW1	SE[d]
Avg. backfat (cm)	2.6	2.4	0.20
Avg. loin eye area (cm^2)	37.4	38.6	0.73
Avg. carcass lean (%)	49.3	50.6	0.70
Avg. loin color score[e]	2.9	2.9	0.08
Avg. loin firmness score[f]	2.7	2.7	0.12
Avg. loin marbling score[g]	2.8	2.4	0.25
Avg. carcass fat firmness score[h]	1.4	1.7	0.16

[a]Each mean is based on the information from four pens of six pigs each. Average weight at slaughter, 112 kg per pig.
[b]Means for each measurement do not differ (*p* > 0.10).
[c]Dehydrated restaurant food waste blended product (blended prior to dehydration); approximately 60% food wastes (dry) and 40% soy hulls/wheat flour blend (55:45); DFW1.
[d]Standard error of the mean; n = 4.
[e]Scores of 1 to 5: 2 = gray, 3 = light pink, 4 = reddish pink.
[f]Scores of 1 to 5: 2 = firm, 3 = slightly firm, 4 = slightly soft.
[g]Scores of 1 to 5: 2 = traces, 3 = slight, 4 = modest.
[h]Scores of 1 to 4: 1 = firm, 2 = slightly soft, 3 = soft, 4 = very soft, oily.

Table 8.12. Carcass characteristics of finishing pigs fed diets containing a dehydrated restaurant food waste product (DFW) trial 2[a]

Item[b]	Dietary treatment (%DFW2[c])			
	0	40	80	SE[d]
Avg. backfat (cm)	2.7	2.5	2.3	0.05
Avg. loin eye area (cm²)	34.8	35.6	34.3	0.42
Avg. carcass lean (%)	48.5	49.8	49.7	0.32
Avg. loin color score[e]	2.6	2.3	2.4	0.10
Avg. loin firmness score[f]	2.5	2.6	2.4	0.06
Avg. loin marbling score[g]	2.3	2.3	2.2	0.10
Avg. carcass fat firmness score[h]	1.5	2.2	2.6	0.12

[a]Each mean is based on the information from three pens of eight pigs each. Average weight at slaughter, 108 kg per pig.
[b]Means for each measurement do not differ ($p > 0.10$) except for backfat ($p < 0.05$; linear) and carcass fat firmness ($p < 0.01$; linear).
[c]Dehydrated restaurant food waste blended product (blended prior to dehydration); approximately 60% food waste (dry) and 40% soy hulls/ground corn blend (67:33); DFW2.
[d]Standard error; n = 4.
[e]Scores of 1 to 5: 2 = gray, 3 = light pink, 4 = reddish pink.
[f]Scores of 1 to 5: 2 = firm, 3 = slightly firm, 4 = slightly soft.
[g]Scores of 1 to 5: 2 = traces, 3 = slight, 4 = modest.
[h]Scores of 1 to 5: 1 = firm, 2 = slightly soft, 3 = soft, 4 = very soft, oily.

Table 8.13. Sensory evaluations and shear force of broiled loin chops from finishing pigs fed diets containing a dehydrated restaurant food-waste product (DFW) trial 1[a]

Item[b]	Dietary treatment		
	0% DFW1[c]	40% DFW1	SE[d]
Juiciness[e]	4.7	5.1	0.2
Flavor[e]	5.7	5.7	0.1
Tenderness[e]	5.7	5.9	0.2
Off-flavor[f]	5.7	5.7	0.1
Shear force[g] (kg/mm)	3.2	2.9	0.1

[a]Each mean is based on the information from four pens of three pigs each. Average weight at slaughter was 112 kg per pig.
[b]Means do not differ ($p > 0.10$).
[c]Dehydrated restaurant food-waste blended product (blended prior to dehydration); approximately 60% food waste (dry) and 40% soy hulls/wheat flour (55:45); DFW1.
[d]Standard error; n = 4.
[e]Scores of 1 to 8 with the trait increasing with an increase in score.
[f]Scores of 1 to 6 with a less intense off-flavor with an increase in value.
[g]Warner-Bratzler shear force values.

Table 8.14. Sensory evaluations and shear force of broiled loin chops from finishing pigs fed diets containing a dehydrated restaurant food-waste product (DFW) trial 2[a]

| Item[b] | Dietary treatment (%DFW2[c]) | | | |
	0	40	80	SE[d]
Juiciness[e]	5.6	5.6	5.6	0.2
Flavor[e]	5.7	5.7	5.9	0.1
Tenderness[e]	6.3	6.3	6.4	0.2
Off-flavor[f]	5.8	5.8	5.8	0.1
Shear force[g] (kg/mm)	2.3	2.7	2.6	0.3

[a]Each mean is based on the information from three pens of four pigs each. Average weight at slaughter, 108 kg per pig.
[b]Means do not differ ($p > 0.10$).
[c]Dehydrated restaurant food waste blended product (blended prior to dehydration); approximately 60% food waste (dry and 40% soy hulls/ground corn blend (67:33); DFW2.
[d]Standard error; n = 3.
[e]Scores of 1 to 8 with the trait increasing with an increase in score.
[f]Scores of 1 to 6 with a less intense off-flavor with an increase in value.
[g]Warner-Bratzler shear force values (measure of tenderness).

inclusion of the DFW products in the diets also had no effect ($P > 0.10$) on shear force values, a measure of tenderness, of the loin chops.

Apparent Digestibility by Pigs

Apparent digestibility of dry matter, nitrogen (protein), and energy of diets containing a dehydrated food-waste product was also done by the University of Florida (Dollar 1998). The DFW product used in this digestibility experiment was the same DFW product (DFW2) as used in finishing trial 2 reported above. Like in finishing trial 2, the DFW2 product was included in corn/soybean meal-based diets at 0 (control), 40, or 80% of the diet. The trial was a replicated crossover design that involved six crossbred barrows. The pigs were housed individually in stainless steel metabolism cages. Each of the three diets was fed for 1 week with a 3-day adjustment and 4-day total fecal and urine collection period. The pigs averaged 40 kg body weight each at the start of the digestibility trial. The diets analyzed 89.5, 88.0, and 89.7% dry matter; 2.8, 2.9, and 2.8% nitrogen; and 4,390, 4,780, and 5,040 kcal/kg of gross energy, respectively, for the 0, 40, and 80% diets. The DFW product itself analyzed 90.1% dry matter, 3.0% nitrogen, and 5,200 kcal/kg of gross energy.

Apparent digestibility coefficients of the 0, 40, and 80% diets obtained in the digestion trial are shown in table 8.15. Decreases in

Table 8.15. Apparent digestion coefficients by pigs for diets containing a dehydrated restaurant food-waste product (DFW2)[a,b]

	0% DFW2 diet (control)	40% DFW2 diet	80% DFW2 diet
Dry Matter	91.1[d]	83.7[c]	82.8[c]
Nitrogen (Protein)	88.6[d]	79.2[c]	77.3[c]
Energy	90.2[d]	82.9[c]	82.6[c]

[a]Least squares means. Each mean is based on the information from six pigs. Average weight range of pigs is 40 to 60 kg per pig.
[b]All numbers are on a dry matter basis of the diets fed.
[c,d]Means within rows without a common superscript differ ($p < 0.01$).

apparent digestibility by the pig of dry matter, nitrogen, and energy were noted as the level of the DFW2 product increased in the diets (P < 0.01). The greatest decreases were noted between the 0 and 40% diets as differences between the 40 and 80% diets were not statistically different (P > 0.05) for dry matter, nitrogen, or energy digestibility. Nitrogen digestibility is an indicator of protein digestibility, thus a decrease in nitrogen digestibility indicates a decrease in protein digestibility. It should be pointed out that the digestibility coefficients obtained for the 0% (control) diet were higher than typically obtained for corn/soybean meal-based diets. Apparent digestibility coefficients are usually in the range of 83 to 88%. Even compared to coefficients typically obtained, the coefficients obtained for the DFW-containing diets indicated less than optimal digestibility of the DFW2 product. The lowered digestibilities were probably the result of the relatively high fiber content of the DFW2 product (table 8.8) and high temperatures encountered during dehydration. The high fiber content was the result of the soy hulls used as part of the feedstock. Fiber is not digested well by the pig, and fiber is also known to have depressing effects on apparent digestibility of other dietary components such as protein (NRC 1988). The high temperature encountered during dehydration, as alluded to above, is also known to reduce digestibility, in particular, protein digestibility (Hurrel 1990; Zhang and Parsons 1996).

In spite of the lowered digestibility values of the DFW2 diets, the data obtained indicated that the pigs were able to effectively utilize most of the energy and protein that this product contributed to the diet. From the digestibility coefficients of the DFW2 diets, the estimated coefficients for the DFW product (DFW2) itself would be in the range of 75 to 80% for dry matter and energy, and 70 to 75% for nitrogen (protein).

Evaluation of a Commercial Dehydrated Food-Waste Product

Another feeding trial with finishing pigs was conducted by the University of Florida on the evaluation of a dehydrated restaurant food-waste product as a feed ingredient for mixed pig diets (Myer et al. 1998). Unlike the previous finishing trials mentioned above, this trial utilized a dehydrated restaurant food-waste product that was developed for the commercial market. The dehydrated product utilized was produced much like that mentioned above but used wheat middlings, a coproduct of flour milling, as the feedstock. In addition, this product was passed through the drying process only once instead of multiple times. As such, the concentration of food waste (dried basis) was estimated to be just 25% with the other 75% being the wheat middlings. This dehydrated product will be referred to as dehydrated wheat middling food-waste blended product (DMFW). After much testing, NutraFeed Inc. determined that the most cost-effective and simplest means of dehydration was with wheat middlings as the feedstock. Wheat middlings have advantages over many other feedstocks in that they are readily available, can be used directly without further processing (i.e., grinding), and produce a pellet that dries easily. On the other hand, even though wheat middlings is a fairly nutritious feedstuff, middlings are rather bulky and high in fiber, which limits their use in pig diets. The maximum level of middlings that can be included in the diet of finishing pigs is normally recommended not to exceed 20% (Holden and Zimmerman 1991).

The finishing trial conducted involved a simple comparison of nutritionally adequate, corn/soybean meal-based diets with or without the DMFW product. The level of DMFW in the experimental diets was 60%. The control diets contained wheat middlings added at the same level as the estimated level of middlings in the 60% DMFW diets; this level was 45%. The nutritionally adequate experimental diets were formulated to be similar in estimated digestible lysine to ME ratio within diet type (composition presented in table 8.16). The pigs were fed the experimental diets from 59 to 113 kg average body weight.

The DMFW product, like the DFW products used previously, was a pelleted, tan-colored product that was slightly greasy to the touch. Unlike that noted before, the color was lighter indicating less exposure to the high temperatures involved in the drying process. This was expected as the product was passed through only once, whereas before, the DFW products were passed through the drying process up to three times. The DMFW pellets were reground before mixing into the finishing diets.

Table 8.16. Percentage composition of experimental diets used in finishing trials with DMFW

Ingredient	Finisher I diets (55 to 80 kg)[a]		Finisher II diets (80 to 110 kg)[a]	
	Control	DMFW	Control	DMFW
DMFW[b]	—	60.00	—	60.00
Wheat middlings	45.00	—	45.00	—
Ground corn	39.25	27.45	47.25	34.40
Soybean meal (48%)	13.00	10.00	5.00	3.00
L-lysine HCl	0.10	0.20	0.10	0.25
Other[c]	2.65	2.35	2.65	2.35
	100.00	100.00	100.00	100.00
Calculated composition[d]				
Crude protein	16.00	18.00	14.00	15.20
Lysine	0.84	0.91	0.62	0.72
Crude fat	3.50	6.00	3.50	6.00
ME (kcal/kg)	3100.00	(3300.00)[e]	3100.00	(3300.00)
ME/lysine (kcal/%)	3700.00	(3600.00)	5000.00	(4700.00)
Analyzed composition[f]				
Moisture	11.50	7.30	12.00	8.20
Crude protein	16.00	17.40	13.60	15.00
Crude fat	3.00	6.40	3.20	6.60
Crude fiber	5.80	6.60	5.60	6.40
Ash	5.50	5.70	5.00	5.60

[a]Pig weight range for which the diets were fed (average per pig).
[b]Dehydrated wheat middling/restaurant food-waste blended product (blended prior to dehydration); contained approximately 25% dried food waste and 75% wheat middlings.
[c]Dietary levels of 0.6 and 0.6% dicalcium phosphate, 1.4 and 1.2% calcium carbonate, 0.3 and 0.1% salt, 0.15 and 0.15% vitamin premix, 0.05 and 0.05% trace mineral premix, 0.15 and 0.15% antibiotic premix, and 0 and 0.1% sodium bicarbonate for control and DMFW diets, respectively. Vitamin premix provided per kg of diet: vitamin A, 3300 IU; vitamin D3, 412 IU; vitamin E, 16 IU; vitamin K activity, 3.3 mg; riboflavin, 4.1 mg; d-pantothenic acid, 14 mg; niacin, 20 mg; choline chloride, 80 mg; and vitamin B12, 16 μg. Trace mineral premix provided per kg of diet: zinc, 100 mg; iron, 50 mg; manganese, 27 mg; copper, 5 mg; iodine, 0.8 mg; and selenium, 0.15 mg. Antibiotic premix provided per kg of diet: tylosin, 55 mg.
[d]Calculated using NRC table values (except lysine in SBM = 3.0%) and values given in table 8.17.
[e]Estimated.
[f]Average of analyses of two samples per diet.

Composition of the major chemical components of the DMFW product is given in table 8.17. Samples of wheat middlings used also were obtained and analyzed with results presented in table 8.17. Since the DMFW was about 75% (dry basis) wheat middlings, the composition reflected this high concentration of wheat middlings. Nevertheless, the DMFW product contained a slightly higher content of crude protein

Table 8.17. Composition and in vitro quality assessments of dehydrated restaurant food waste/wheat middlings blended product (DMFW) and of wheat middlings used in pig finishing trial with DMFW[a]

Item	Middlings[b]	DMFW[c]
Moisture (%)	13.20	6.50
Crude protein (%)	17.00	18.60
Lysine (%)	0.62	0.63
Available lysine (%)	0.56	0.55
Threonine (%)	0.53	0.62
Isoleucine (%)	0.38	0.48
Pepsin indigestible protein (%)	2.60	3.50
Crude fat (%)	3.60	8.60
Crude fiber (%)	8.80	9.60
Total mineral matter (%)	4.80	5.20
Calcium (%)	0.10	0.34
Phosphorous (%)	1.00	1.00
Chloride-soluble (%)	0.01	0.23
Sodium (%)	0.07	0.36

[a]As fed basis.
[b]Same middlings as used in the dehydration of DMFW and in the control diet of the finishing trial.
[c]Contained approximately 25% food waste (dry basis) and 75% wheat middlings.

and more than double the content of crude fat compared to the wheat middlings alone. Amino acid analysis, however, indicated that the DMFW product contained only slightly higher content of the important essential amino acids (lysine, threonine, isoleucine, methionine) than the wheat middlings. Furthermore, *in vitro* pepsin digestibility and lysine availability assays indicated that the protein value of the DMFW product was essentially the same as that of the wheat middlings even though the DMFW product contained slightly more protein than the middlings. Heat processing, like that encountered in the dehydration process for DMFW, is known to decrease the digestibility of protein and of individual amino acids (Hurrell 1990; Zhang and Parsons 1996). Heat processing also has been shown to decrease analyzed contents of certain amino acids such as lysine (Zhang and Parsons 1996). The DMFW product contained an appreciable amount of salt (estimated to be 0.75%) in spite of the high content of wheat middlings.

Growth performance results of the pig finishing trial are summarized in table 8.18. The average rate of weight gain while the pigs were on test was similar (P > 0.10) for pigs fed the 60% DMFW experimental diets to that of pigs fed the 45% wheat middlings control diets. The amount of feed required per unit of weight gain for pigs fed the 60% DMFW diets was 9% less (P < 0.01) than pigs fed the control diets. This improvement was probably the result of the higher fat content, contributed by the

Table 8.18. Performance and carcass characteristics of finishing pigs fed diets containing a dehydrated wheat middlings/restaurant food waste product (DMFW) trial 3[a]

| | Dietary treatment | | |
Item	Control (45% wheat middlings)	60% DMFW[b]	SE[c]
Avg. daily gain (kg)	0.896	0.892	0.015
Avg. daily feed intake[d] (kg)	3.270	3.010	0.037
Feed/unit gain[d] (kg/kg)	3.650	3.310	0.051
Avg. backfat (cm)	1.800	1.700	0.070
Avg. carcass lean[e] (%)	50.400	49.700	0.550

[a]Five pens per treatment with six pigs per pen. On experiment from 59 to 113 kg average body weight per pig.
[b]Dehydrated restaurant food waste/wheat middlings blended product (blended prior to dehydration); approximately 25% food waste (dry) and 75% wheat middlings; DMFW.
[c]Standard error; n = 5.
[d]Means differ ($p < 0.01$).
[e]Fat free lean.

DMFW, to the 60% DMFW diets compared to the control diets (table 8.16). Carcass lean content for the pigs, as estimated by real time ultrasound at the end of the trial, was not affected by dietary treatment (P > 0.10; table 8.18).

The results of this trial concur with those of the previous two finishing pig trials mentioned earlier in this chapter in that dehydrated foodwaste products are nutritious feedstuffs for inclusion in mixed pigs' diets. Based on the improvement in feed conversion efficiency noted in the third finishing trial and its higher fat content, the energy value of the DMFW product was estimated to be 15% greater than just the wheat middlings. However, even though the DMFW product contained more analyzed protein than the wheat middlings, its protein value would only be equal to wheat middlings based on the similar amounts of available protein and lysine.

Summary

Overall, DFW products evaluated in the above Florida trials had high contents of protein and fat that are desirable for pig feeding. It appears that most of this protein and fat can be utilized by the pig. Microbiological evaluations conducted in the initial trial indicated normal levels and types of microbes present in the gut of pigs consuming diets containing DFW and thus should pose no particular health threat to pigs or humans. The slight negative effects noted with protein digestibility and fat rancidity can be minimized by proper temperature selection during drying, minimizing drying time, use of an antioxidant, and by the preparation of food waste before drying (i.e., selection of feedstock, ensuring that the waste is fresh, etc.).

The results of the initial three finishing pig trials indicated the DFW products used were nutritious feedstuffs for inclusion in pig diets. Feed intake by pigs fed diets containing these products was similar, on an estimated calorie basis, as the control-fed pigs. The actual amount of dehydrated food waste in the diets, minus feedstocks, was estimated to be 25 and 50% when DFW product was included at 40 or 80% of the diet for the first two trials, respectively. In the third trial, the actual amount was 15% when 60% DMFW was included in the diet. The improvement in feed efficiency when pigs were fed diets containing DFW, compared to the control diet in each trial, indicated that the DFW products with their higher fat content were well-used by the pigs. The softening effect on carcass fat noted in the second finishing trial may limit the amount of DFW products in the diet. Furthermore, the DFW products utilized were easy to further process and mix into the meal-type diets utilized in the pig finishing trials. The diets flowed easily through the self-feeders.

In the above finishing trials, the dehydrated food-waste products were part of mixed diets. To maximize the benefit for pigs of what could be contributed to the diet (i.e., fat) and to minimize possible diet problems (i.e., high salt), it would be best that DFW products be part of a mixed diet. The maximum diet inclusion levels would depend on the concentration of dried food waste in the final DFW product and/or type of feedstock used. For example, in finishing trial 2, the DFW product used should be limited to 50% of the diet or less to minimize the softening effect on the carcass fat. In trial 3, the wheat middlings in the DFW product would limit the amount that can be included in the diet.

Feeding a Dehydrated Food-Waste Product to Sows

Due to the higher fiber content that can be associated with the dehydrated food waste (DFW) products, it was felt that a DFW product might be more appropriate for inclusion into diets for breeding swine. Therefore, trials that involved gestating and lactating sows were also done at the University of Florida.

The first trial was done with gestating sows that were fed DFW-containing diets to determine the resulting effects on sow weight change, and on the number born and birth weight of pigs in the subsequent litter (Dollar 1998). Two different batches of DFW were used in the gestation diets. Batch 1 contained approximately 48% restaurant food waste

Table 8.19. Chemical composition (%) of corn, soybean meal, and dehydrated restaurant food waste products (DFW-P)[a] used in the sow trials[b]

Item	DFW-P (batch 1)	DFW-P (batch 2)	Corn	Soybean meal
Dry matter	93.0	91.6	86.6	88.8
Ash[c]	5.5	12.1	1.4	7.3
Gross energy (kcal/kg)	5185.0	4665.0	4225.0	4545.0
Fat	14.9	10.0	4.6	2.7
Protein	15.5	13.3	8.9	48.0
Calcium	1.0	2.3	0.0	0.4
Phosphorus	0.9	0.5	0.3	0.8
Sodium	0.5	0.3	0.0	0.0
Total neutral detergent fiber	36.7	45.6	7.9	9.0
Ash free N.D.F.	35.9	43.3	7.0	8.1

[a]Dehydrated resturant food waste blended product; approximately 48% food waste (dried) and 52% feedstock (60:40 peanut hulls:wheat middlings) - batch 1, or 28% food waste and 72% feedstock (60:40 peanut hulls:wheat middlings) - batch 2.
[b]Values are expressed on a dry matter basis.
[c]Total mineral matter.

Table 8.20. Composition (%) of diets fed during gestation of the sow trials[a]

Item	Control diet	DFW-P diet 1	DFW-P diet 2
Corn	73.200	0.000	0.000
Soybean meal	15.000	0.000	5.400
DFW-P[b]	—	97.600	92.200
Corn oil	7.500	0.000	0.000
Iodized salt	1.000	0.000	0.000
Trace minerals[c]	0.075	0.075	0.075
Vitamin premix[d]	0.100	0.100	0.100
Dynafos®[e]	2.500	2.200	2.200
Limestone	0.600	0.000	0.000
Lysine HCl	0.025	0.025	0.050
Calculated composition			
Lysine	0.650	0.650	0.650
Calcium	0.900	0.900	0.900
Phosphorus	0.750	0.750	0.750
Sodium	0.450	0.450	0.450
ME (mcal/kg)	3.500	(3.500)[f]	(3.500)

[a]Values are expressed on a percent as-fed basis.
[b]Dehydrated restaurant food waste products (batch 1, 48% dried food wastes and 52% feedstock; batch 2, 28% dried food waste and 72% dry feedstock).
[c]Provided per kg of diet: 200 mg of zinc, 100 mg of iron, 55 mg of manganese, 11 mg of copper, 1.5 mg of iodine, 1 mg of cobalt, and 20 mg of calcium.
[d]Provides per kg of diet: 13.2 mg of riboflavin, 44 mg of niacin, 26.4 mg of d-pantothenic acid, 176 mg of choline, 22 μg of vitamin B$_{12}$, 5,500 IU of vitamin A, 880 IU of vitamin D$_3$, 22 IU of vitamin E, and 3 mg of vitamin K.
[e]Dynafos is a registered product of IMC Agrico, Bannockburn, IL, and is a chemical mixture of monocalcium and dicalcium phosphates.
[f]Estimate.

(dry basis) with 52% dry feedstock. Batch 2 contained approximately 28% restaurant food waste (dry basis) with 72% dry feedstock. The dry feedstock used was mainly a mixture of peanut hulls and wheat middlings (about 60:40 mix). Chemical composition of the two DFW (DFW-P) batches is found in table 8.19. Two gestation diets, each containing a different batch of DFW-P product were formulated and compared to a control diet (dietary compositions presented in table 8.20). Diets were fed from breeding to farrowing at a rate of 2.5 kg/day for sows consuming the control and DFW-P diet 1, and those consuming DFWP diet 2 were fed 3.6 kg/day due to differences in dietary bulk density.

Weight changes during the first 30 days of gestation were extremely variable, probably because sows were assigned to diets after weaning of the previous litter and gilts at the time of breeding. Thus, weight changes recorded represented day 30 to 110 of gestation (data presented in table 8.21). Sows consuming the control diet gained more weight (P < 0.05) during days 30 to 60 of gestation than sows fed either DFW-P diet 1 or 2. Sows fed the DFW-P diets gained similarly during this time period. No effect (P > 0.10) on sow weight was noted from day 60 to 90 of gestation due to dietary treatments. When considering weight gain from day 30 to 110, sows fed the control diet gained more weight (P < 0.01) than sows consuming either of the DFW-P diets. Similar weight gain was noted between sows fed the DFW-P diets. Metabolizable energy content between the control and the DFW-P diets was probably the reason for the difference in weight gain. The dry feedstock used in the DFW-P product contained a high level of peanut hulls that have a much lower energy digestibility (NRC 1982) as compared to corn and soybean meal. Therefore, even though diets were similar in gross energy content, digestible and metabolizable energy contents would have

Table 8.21. Effects of feeding diets containing dehydrated food waste product (DFW-P) during gestation on sow and litter performance[a]

	Control diet	DFW-P diet 1	DFW-P diet 2	CV[b]
Gestation wt. change day 30 to 60 (kg)	20.70[c]	9.40[d]	11.50[d]	70.1
Gestation wt. change day 60 to 90 (kg)	17.20[c]	13.50[c]	13.50[c]	64.1
Gestation wt. change day 30 to 110 (kg)	59.80[c]	28.80[d]	34.50[d]	50.7
Number born	10.00[c]	11.60[cd]	12.50[d]	24.4
Litter birth wt (kg)	15.10[c]	16.50[c]	17.10[c]	24.8
Average pig birth wt (kg)	1.51[c]	1.42[cd]	1.35[d]	19.5
Number of observations	10.00	16.00	32.00	

[a] Least squares means.
[b] Coefficient of variation.
[cd] Means within rows without a common superscript differ ($p < 0.10$).

differed. Although sows fed the DFW-P diets gained less weight in gestation, they produced more pigs (P < 0.10) than sows fed the control diet. This greater number of pigs, however, did not result in heavier total litter birth weights (P > 0.10) as average pig birth weight was slightly lower (P < 0.10) from sows fed the DFW-P diets.

In addition to feeding the DFW-P during gestation, two trials were conducted using DFW-P in lactation diets (Dollar 1998). It was felt that the increase in diet bulk associated with DFW-P product would be beneficial in preventing constipation in the lactating sow. It is common for sows confined to farrowing crates to become constipated due to a lack of exercise, reduced water consumption, consumption of lactation diets with a low fiber content, and physiological and hormonal changes that occur at the time of parturition (Shurson 1993). Dietary fiber is relatively indigestible for the pig and acts as a bulking agent absorbing water and preventing the occurrences of dry and hardened feces. However, too much fiber in the diet increases the bulk density and may cause the sow to consume inadequate calories. This reduction in caloric intake can result in excessive sow weight loss resulting in poorer milk production and delayed return to estrus (Shurson 1993). Thus the hypothesis was to feed DFW-P product as a partial dietary replacement so as not to lower energy density appreciably. Therefore in trial 1 of lactation, DFW-P was added at 52% of the diet and the DFW-P product used was from batch 1 (table 8.19). Trial 2 used DFW-P from batch 2 (table 8.19) and was included at 27% in the diet. Compositions of the lactation diets are presented in tables 8.22 and 8.23. Less DFW-P was used in trial 2 because batch 2 of the DFW-P contained a much higher level of neutral detergent fiber from the higher porportion of feedstock (mostly peanut hulls) and thus would have a lower energy content.

Sows were brought in the farrowing house at approximately day 110 of gestation. Sows were fed a gestation diet until the time of parturition and were placed on their respective lactation dietary treatments following parturition. Sow weight change was measured from farrowing to weaning, which occurred at 21 days. Sow feed intake was recorded also for the entire lactation, and pig weight gain from birth to weaning was noted as well as number of pigs weaned and litter weaning weight.

Sows that were fed the control diet in trial 1 tended to produce pigs that gained more weight during lactation than the pigs from sows consuming the DFW-P product-containing diets. However, this effect was significant only in trial 2 in which sows fed the control diet produced pigs that gained more weight (P < 0.07) than pigs nursing sows consuming the DFW-P containing diet. This was the only litter criterion measured that was significantly different between the dietary treatments for both

Table 8.22. Composition (%) of diets for sow lactation trial 1[a]

Item	Control diet	DFW-P diet
Corn	57.600	29.500
Soybean meal	23.600	16.100
DFW-P[b]	0.000	52.000
Corn oil	7.700	0.000
Wheat bran	7.500	0.000
Iodized salt	0.500	0.000
Trace minerals[c]	0.075	0.075
Vitamin premix[d]	0.100	0.100
ASP-250	0.150	0.150
Dynafos®[e]	2.000	2.100
Limestone	0.780	0.100
Calculated composition		
Lysine	0.900	0.900
Calcium	0.850	0.850
Phosphorus	0.750	0.750
ME, (mcal/kg)	3.500	(3.500)[f]

[a]Values are expressed on an as-fed basis.
[b]Dehydrated restaurant food-waste product (batch 1).
[c]Provides per kg of diet: 200 mg of zinc, 100 mg of iron, 55 mg of manganese, 11 mg of copper, 1.5 mg of iodine, 1 mg of cobalt, and 20 mg of calcium.
[d]Provided per kg of diet: 13.2 mg of riboflavin, 44 mg of niacin, 26.4 mg of d-pantothenic acid, 176 mg of choline, 22 μg of vitamin B_{12}, 5,500 IU of vitamin A, 880 IU of vitamin D_3, 22 IU of vitamin E, and 3 mg of vitamin K.
[e]Dynafos is a registered product of IMC Agrico, Bannockburn, IL, and is a chemical mixture of monocalcium and dicalcium phosphates.
[f]Estimated.

trials. The other criteria, number of pigs weaned and litter weaning weight, were not different (P > 0.10) between dietary treatment for either trial (data presented in tables 8.24 and 8.25). There was some evidence, however, that sows fed the control diet provided more milk since litter weaning weights tended to be heavier in both trials when sows were fed the control diet versus the DFW-P containing diets. This again was due most probably to a lower metabolizable energy content in the diets containing DFW-P. The neutral detergent fiber levels were 28.6 and 23.1% for the DFW-P containing diets in trials 1 and 2, respectively, while the control diets contained only 13.1 and 12.5% neutral detergent fiber in trials 1 and 2, respectively. Since feed intake was not different between dietary treatments or between trials, it is most probable that sows consuming the diets with DFW-P consumed less available energy and thus produced less milk and therefore slightly lower litter weaning weights. The lower available energy in the diets containing DFW-P is further evidenced when one considers the sow weight change in trial 2. Although sows fed the DFW-P diet ate the same amount of feed as the

Table 8.23.   Composition (%) of diets for sow lactation trial 2[a]

Item	Control diet	DFW-P diet
Corn	61.200	48.400
Soybean meal	23.600	21.500
DFW-P[b]	0.000	27.000
Corn oil	4.000	0.000
Wheat bran	7.500	0.000
Iodized salt	0.500	0.300
Trace minerals[c]	0.075	0.075
Vitamin premix[d]	0.100	0.100
ASP-250	0.150	0.150
Dynafos7[e]	2.400	2.200
Limestone	0.550	0.330
Calculated composition:		
Lysine	0.900	0.900
Calcium	0.850	0.850
Phosphorus	0.750	0.750
ME (mcal/kg)	3.300	(3.300)[f]

[a]Values are expressed on an as-fed basis.
[b]Dehydrated food waste product (batch 2).
[c]Provided per kg of diet: 200 mg of zinc, 100 mg of iron, 55 mg of manganese, 11 mg of copper, 1.5 mg of iodine, 1 mg of cobalt, and 20 mg of calcium.
[d]Provides per kg of diet: 13.2 mg of riboflavin, 44 mg of niacin, 26.4 mg of d-pantothenic acid, 176 mg of choline, 22 μg of vitamin B_{12}, 5,500 IU of vitamin A, 880 IU of vitamin D_3, 22 IU of vitamin E, and 3 mg of vitamin K.
[e]Dynafos is a registered product of IMC Agrico, Bannockburn, IL, and is a chemical mixture of monocalcium and dicalcium phosphates.
[f]Estimated.

Table 8.24. Effects of lactation dietary treatment on sow and litter performance during a 21-day lactation (trial 1)[a]

	Control diet	DFW-P[b] diet	SE[c]
Pig weight gain (kg)	4.2[d]	4.0[d]	0.2
Number weaned	10.2[d]	10.0[d]	0.5
Litter weaning weight (kg)	55.8[d]	53.7[d]	2.9
Sow feed intake (kg)	115.0[d]	118.5[d]	3.0
Sow weight change (kg)	2.9[d]	4.9[d]	2.5
Days to estrus	3.8[d]	4.2[d]	0.3
Number of observations	20	18	

[a] Least squares means.
[b] Dehydrated restaurant food waste product (batch 1).
[c] Standard error.
[d] Means within a row without a common superscript differ ($p < 0.10$).

control-fed sows, they tended to produce lighter litter weights and lost more weight (P < 0.02) during lactation. This change in body weight also may have affected their reproductive state since sows in both trials

Table 8.25.   Effects of lactation dietary treatment on sow and litter performance during a 21-day lactation (trial 2)[a]

	Control diet	DFW-P[b] diet	SE[c]
Pig weight gain (kg)	5.4[d]	4.9[e]	0.2
Number weaned	10.2[d]	10.7[d]	0.4
Litter weaning weight (kg)	69.4[d]	66.4[d]	2.6
Sow feed intake (kg)	111.0[d]	112.6[d]	2.6
Sow weight change (kg)	4.8[f]	−2.7[g]	2.2
Days to estrus	4.8[d]	5.1[d]	0.2
Number of observations	26.0	23.0	

[a]Least squares means.
[b]Dehydrated restaurant food waste product (batch 2).
[c]Standard error.
[d,e]Means within a row without a common superscript differ ($p < 0.10$).
[f,g]Means within a row without a common superscript differ ($p < 0.02$).

that were fed the DFW-P diets tended to take longer to return to estrus (data presented in tables 8.24 and 8.25).

Summary of Sow Trials

The research to date conducted with gestating and lactating sows that were fed dehydrated food-waste products was encouraging even though the DFW-P products contained peanut hulls as part of the dry feedstock used in the dehydration process. Although diets containing the DFW-P appeared to have less available energy, because of the peanut hulls, they provided satisfactory sow and litter performance. In future trials, it may be necessary to decrease or eliminate the use of peanut hulls as the dry feedstock. A dry feedstock such as wheat middlings with or without soybean hulls may be a better alternative to blend with the food waste prior to dehydration.

Concluding Remarks

While the research to date has indicated that the dehydration of food waste produces a safe and nutritious feedstuff for finishing pigs and sows, nevertheless, the Swine Health Protection Act (1980) and various state regulations will have to be amended to allow the feeding of DFW like that described in the above research trials. Temperatures utilized during dehydration in the above trials easily exceeded that required by the Swine Health Protection Act; however, the requirement that 100°C be maintained for 30 minutes was not. Depending upon interpretation, DFW may be classified as a rendered product, as was the case by the state of Florida for the above research. Regulations to produce a rendered product were followed in the above research trials.

Conclusion

The dehydration of restaurant food waste can potentially produce a nutritious feedstuff for use in swine diets while also offering a viable solid waste disposal option.

References

Ammerman, C. B. and P. R. Henry. 1991. Citrus and vegetable products for ruminant animals. Proc. Alternative Feeds for Dairy and Beef Cattle Sym. E. R. Jordan, ed. Univ. of Missouri, Columbia.

Blake, J. P., M. E. Cook, and D. R. Reynolds. 1991. Extruding poultry farm mortalities. Am. Soc. Agric. Engr. Internat'l. meeting. Paper No. 914049. Am. Soc. Agric. Engr., St. Joseph, MI.

Dollar, K. K. 1998. The use of dried restaurant food residual products as a feedstuff for swine. M. S. Thesis. Univ. of Florida, Gainesville, FL.

Ferris, D. A., R. A. Flores, C. W. Shanklin, and M. K. Whitworth. 1995. Proximate analysis of food service wastes. *Appl. Engr. Agric.* 11:567-572.

Haque, A. K. M. A., J. J. Lyons and J. M. Vandepopuliere. 1991. Extrusion processing of broiler starter diets containing ground whole hens, poultry by-product meal, feather meal, or ground feathers. *Poul. Sci.* 70:234-240.

Holden, P. J. and D. R. Zimmerman. 1991. Utilization of Cereal Grain By-Products in Feeding Swine. Swine Nutrition, pp. 585-593. E. R. Miller, D. E. Ullrey, and A. J. Lewis, eds. Butterworth-Heinemann. Stoneham, U.K.

Hurrell, R. F. 1990. Influence of the maillard reaction on the nutritional value of foods. The Maillard Reaction in Food Processing, Human Nutrition and Physiology. Birkhanser Verlag, Basel, Switzerland.

Kornegay, E. T., G. W. Van der Noot, K. M. Barth, G. Graber, W. S. MacGrath, R. L. Gilbreath, and F. J. Bielk. 1970. Nutritive evaluation of garbage as a feed for swine. New Jersey Experiment Station. Bull. No. 829. Rutgers, New Brunswick, NJ.

Lyons, J. J. and J. M. Vandepopuliere. 1996. Spent leghorn hens converted into a feedstuff. *J. Appl. Poultry Res.* 5:18-25.

Myer, R. O. 1998. Evaluation of a dehydrated poultry (broiler) mortality—soybean meal product as a potential supplemental protein source for pig diets. Proc. Internat'l Conference on Animal Production Systems and the Environment. Iowa State University, Ames.

Myer, R. O., J. H. Brendemuhl, and D. D. Johnson. 1998. Evaluation of dehydrated restaurant food waste products as feedstuffs for finishing pigs. *J. Anim. Sci.* (in press).

Myer, R. O., T. A. DeBusk, J. H. Brendemuhl, and M. E. Rivas. 1994. Initial assessment of dehydrated edible restaurant waste (DERW) as a potential feedstuff for swine. Florida Swine Res. Rep. No. ANS-SW94. University of Florida, Gainesville.

Myer, R. O., D. D. Johnson, W. S. Otwell, and W. R. Walker. 1987. Evaluation of extruded scallop viscera-soy mixtures in diets for growing-finishing swine. *J. Anim. Sci.* 65 (suppl. 1): 33 (Abstract).

NRC (National Research Council). 1982. United States-Canadian tables of feed composition. 3rd Ed. National Academy Press, Washington, D.C.

NRC. 1988. Nutrient Requirements of Swine. 9th Ed. National Academy Press, Washington, D.C.

Pace, B. 1997. Personal Communication.

Pond, W. G. And J. H. Maner. 1984. Swine Production and Nutrition. Westport, CT: AVI Publishing Co., Inc.

Rivas, M. E., J. H. Brendemuhl, R. O. Myer, and D. D. Johnson. 1995. Chemical composition and digestibility of dehydrated edible restaurant waste (DERW) as a feedstuff for swine. *J. Anim. Sci.* 73 (suppl. 1): 177 (Abstract).

Shurson, J. 1993. Research examines pre-farrowing, post-lactation feeding. Feedstuffs. 65:12.

Tacon, A. G. J. and A. J. Jackson. 1985. Utilization of conventional and unconventional protein sources in practical fish feeds. Nutrition and Feeding in Fish. (C. B. Cowey, A. M. Mackie and J. B. Bell, Eds.). New York: Academic Press. pp. 118-145.

Tadtiyanant, C., J. J. Lyons and J. M. Vandepopuliere. 1993. Extrusion processing used to convert dead poultry, feathers, eggshells, hatchery waste, and mechanically deboned residue into feedstuffs for poultry. *Poul. Sci.* 72:1515-1527.

Walker, P. M. and T. Kelly. 1997. Selected fractionate composition and microbiological analysis of institutional food waste, pre- and postextrusion. Proc. Food Waste Recycling Sym. Rutgers Coop. Ext., Rutgers Univ. - Cook College, New Brunswick, NJ.

Walker, P. M. and A. E. Wertz. 1994. Analysis of selected fractionates of a pulped food waste and dish water slurry combination collected from university cafeterias. *J. Anim. Sci.* 72 (suppl. 1): 137 (abstract).

West, R. L. and R. O. Myer. 1987. Carcass and meat quality characteristics of swine as affected by the consumption of peanuts remaining in the field after harvest. *J. Anim. Sci.* 65:475- 480.

Westendorf, M. L. 1996. The use of food waste as a feedstuff in swine diets. Proc. Food Waste Recycling Sym. Rutgers Coop. Ext., Rutgers Univ. - Cook College, New Brunswick, NJ.

Westendorf, M. L., E. W. Zirkle, and R. Gordon. 1996. Feeding food or table waste to livestock. *The Professional Anim. Scien.* 12:129-137.

Zhang, Y. and C. M. Parsons. 1996. Effects of overprocessing on the nutritional quality of peanut meal. *Poul. Sci.* 75:514-518.

9

Case Studies in Utilizing Food-processing By-products as Cattle and Hog Feed

by H. W. Harpster

Introduction

Overview

Agricultural by-product or nonconventional feeds fall largely into one of four areas: food-processing residues, crop residues, forest product residues, and animal waste. Food-processing residues are abundant in many areas and represent the most versatile class of by-product feeds in terms of suitability for both ruminants and nonruminants.

The food-processing industry is critical to the economic well-being of many regions while generating enormous quantities of plant and animal-based by-products. Much of this waste can be safely and profitably fed to livestock but an organized source of nutritional management information is not available and limits producer use of these materials. Using Pennsylvania as an example, an excellent summary of the current scope of the waste disposal problem can be found in Brandt and Martin (1994).

In addition to the justification for extending by-product feeding from the livestock producer's perspective, there are further considerations with regard to environmental quality and industrial employment. Agriculture is a major contributor of waste materials that are increasingly presenting serious disposal problems. Processing plants, often located in or near urban centers, face increasing waste disposal pressures. As environmental legislation becomes more rigid, plant relocation or shutdown becomes a distinct possibility, adversely affecting the economic well-being of the entire area. At the same time, a farmer successfully feeding by-products can extend markedly the animal production potential of his or her farm without expanding the land base of the farm. The future of

145

Table 9.1. Pennsylvania food processing industries

Type	Number
Meat packing plants	80
Sausage and other prepared meats	82
Poultry slaughter/processing	26
Cheese, natural and processed	24
Ice cream, frozen desserts	35
Fluid milk	76
Canned specialties	6
Canned fruit and vegetables	25
Bread, cake, and related products	155
Cookies and crackers	45
Malt beverages	12
Bottled and canned soft drinks	66

Source: U.S. Census of Manufacturers 1987.

our agricultural economy is dependent upon our ability to formulate sound management strategies for dealing with the ever-increasing amounts of waste materials generated. When successfully formulated, these strategies benefit the farmer, agribusiness companies and their employees, and the environment.

Potential

An impressive number and variety of food-processing plants exists in many states. In Pennsylvania alone, more then 600 processing plants have been identified (table 9.1).

Obviously, not all industries produce a by-product that can be fed to livestock, but many options exist. To identify those materials offering potential as feedstuffs, one should attempt to answer the following questions:

• Does the waste have potential nutritional value for livestock? The protein content and estimated energy value of the material must be determined. In addition, the economic value of the waste product will depend in part upon the replacement value the material has for a conventional feed source.
• Is the waste likely to be palatable and acceptable to the animals, thereby allowing a high consumption of the total ration containing the waste material? If decreased intake of dry matter occurs with the waste material, it may not have a true economic value to the feeding program even if nutrient density is adequate.
• Does the waste contain metals, plastic, or other contaminants? One of the first considerations made in preparing a waste material for feeding is to make certain that these kinds of contaminants are not in the waste product stream.

- Does the waste contain toxic materials or chemical residues? Heavy metal contamination may occur naturally in some food-processed waste materials; these need to be evaluated because they affect the health and productivity of all livestock.
- What is the amount of the waste material and seasonality of production? The amount of material produced that is recycled as a potential feed has to be supplied in sufficient quantities and the consistency of product needs to be assessed. If the waste has potential for ensiling, then it can be incorporated into the total mixed ration over a longer period of time.
- What are the handling and storage considerations? Many waste products have a relatively high moisture level, and transportation, handling, and storage can result in problems to both the processor and the producer. Therefore, both must work together in understanding each other's restrictions in handling such waste. The additional costs of transportation, storage, and handling must be included in arriving at a fair market value.
- What is the current level of demand for the material? In some cases, the food processor is willing to pay the livestock producer for removing the waste. However, such arrangements are typically based on relatively long-term contracts, and the producer must be equipped to dispose of all the processing waste generated.

Economic Considerations

The degree to which by-products will be used in farm animal diets is obviously dependent on the economic value of the by-products. This is highly related to the market price of conventional feedstuffs. In our experience, the amounts and types of by-products used by livestock producers changes with the corn and soybean markets. Yet establishing a fair market price for a by-product involves consideration of many factors under constantly changing conditions.

It has been apparent over the years that by-product ingredients capable of entering the mainstream feed trade quickly become established, "conventional" feedstuffs. As such, the price of these ingredients usually rises as demand increases. A summary of characteristics related to the degree of risk and profit potential (from the livestock producer's perspective) of by-product feeds is presented in table 9.2.

Many of the feedstuffs listed in the low risk and profit categories are now purchased by brokers and processed into blends that are sold to commercial feed mills. Other ingredients are incorporated into pet food, often at prices above what is justified for farm animals.

High-moisture by-products that are characteristic of the high-risk (but also high-profit potential) category provide challenging nutritional

Table 9.2. Factors affecting the risk of use and profit potential of by-product feeds for livestock producers

Factor	Risk and Profit Potential	
	Low	High
Moisture	Dry	Wet
Nutritive value	High	Low-medium
Contamination potential	Low	High
Shelf life	Stable	Perishable
Transport	Easy/economical	Difficult/expensive
Storage	Multiple options	Limited options
Processing	Simple	Complex
Availability	Continuous	Seasonal
Market channels	Established	Limited
Examples	Pasta	Vegetable trimmings
	Bakery waste	Cannery waste
	Candy waste	Pomaces
	Rendered proteins	Table waste
	Grain by-products	Liquids

Table 9.3. Relative value of certain waste materials and standard feeds at different distances from the processing plant

Feed	Dry Matter (%)	Relative Value-Distance from Plant		
		At Plant	5 Miles	50 Miles
Corn silage	27.6	100	66	51
Citrus pulp	22.8	127	87	72
Citrus pulp	90.0	501	385	371
Corn cannery waste	22.4	93	61	45
Grape waste	30.0	41	19	4
Apple waste	30.0	115	78	63
Tomato waste	11.1	45	22	7
Pea cannery waste	24.5	86	55	40
Potato waste	20.0	72	66	29

management problems. A few examples of relative economic value are presented in table 9.3 (Wilson and Lemieux 1980).

In this example, the value of field-harvested, whole plant corn silage was set at 100% and the other feedstuffs compared to it. Trucking and handling by conventional farm trucks and systems were assumed in the calculations. Notice the tremendous impact on feed value from moisture content and distance from the source of by-product to the point of feeding.

Other approaches to calculating the economic value of a variety of by-products can be found in Grasser et al. (1995), Rogers and Poore (1994), and Eastridge (1995a).

Case Studies

There are literally hundreds of underutilized materials that could be reviewed, as evidenced by a popular database of by-products and unusual feeds assembled annually by Bath et al. (1997). Further, an ever-increasing number of recent symposia on the subject of by-product feeds is in evidence at both the regional (Westendorf and Zirkle 1996) and national (Eastridge 1995b) levels and provides specific new information on a wide variety of materials from food-plate waste to sea clam-processing residuals to liquid supplements.

Rather than attempt a review of the many by-product feed options available, a case study approach will be presented based on four products from three diverse food industries. Each was targeted to a specific farm animal situation and required a different management scheme for adequate utilization of the resource. Three of the studies were completed on the Penn State campus and one was conducted on a private swine farm.

Case Study 1

Problem. A large manufacturer of human liquid supplements was disposing of huge quantities of infant formula, most in canned form, by landfill at a tremendous economic loss. To ensure consumer confidence, the manufacturer removes such products from the shelf after a relatively short period of time. Nutrient analysis and ingredient information indicated that, with the exception of suboptional levels of protein, the formula appeared desirable for young calves, perhaps substituting for commercial milk replacer in veal, replacement heifer, or calf backgrounding programs. It was obvious that the canned formulas maintained suitable quality for animals well beyond the allowable shelf life for human consumption. In addition, the product could be obtained for the cost of shipping or in some situations at no cost.

Approach. A study was designed (Swope et al. 1995) involving 30 individually fed male Holstein calves obtained shortly after birth. Ten calves were randomly assigned to each of three dietary treatments (table 9.4).

Body weights were taken, and diet volumes were adjusted weekly. Diets were increased from 12% DM and 10% of body weight at week 1 to 14% DM and 12% of body weight at week 7. Calves were weaned to dry feed during week 7 and fed ad lib through week 10. Blood samples were collected at week 0, 6, and 10. The data were analyzed as a split plot in time experiment.

Results. Performance and blood parameters are presented for the 6-week liquid feeding phase only (table 9.5).

Table 9.4. Diet composition (DM basis) of three liquid feeds for young calves

Item	Treatment[a]		
	CMR	IFS	IFW
Dry matter (%)	14.0	14.0	14.0
Crude protein (%)	23.0	23.0	23.0
Crude fat (%)	12.0	28.9	19.5
Calcium (%)	0.8	0.4	0.5
Phosphorus (%)	0.6	0.4	0.5
Potassium (%)	1.9	1.0	1.2
Magnesium (%)	0.2	0.1	0.1
Iron (ug/g)	37.0	57.0	14.0

[a]CMR = commercial milk replacer; IFS = infant formula + Promocaf® (soy protein concentrate); IFW = infant formula plus whey protein concentrate

Table 9.5. Performance of calves fed three liquid milk replacer diets

Item[a]	Treatment		
	CMR	IFS	IFW
DMI/d,kg	0.88[b]	0.80[c]	0.89[b]
ADG,kg	0.70[b]	0.45[c]	0.73[b]
G/F	0.79[b]	0.56[c]	0.82[b]
Hemoglobin, g/dl	10.86	10.37	10.96
Packed cell volume, %	33.17	31.17	33.40
Glucose, mg/dl	113.00	130.00	125.00
Insulin, ng/ml	7.14	4.46	6.04
Urea nitrogen, mg/dl	5.60[b]	8.00[c]	4.90[b]

[a]DMI = Dry Matter Intake; ADG = Average Daily Gain; G/F = Gain to Feed Ratio.
[b,c] means without a common superscript differ ($p < .01$).

When outdated infant formula was supplemented with a high-quality protein like whey (IFW), performance was equal to the commercial milk replacer. Mean blood parameters were similar with the exception of plasma urea nitrogen, which was elevated in calves receiving infant formula supplemented with soy protein (IFS).

Implications. Outdated infant formula, when supplemented with a high quality protein, can be used in calf-rearing systems successfully. Although performance was depressed when a soy-based protein was used to supplement the infant formula, feed costs were dramatically reduced. On a kilogram of live weight gain basis, feed costs were $2.21 for the commercial milk replacer (CMR), $0.55 for the IFS treatment, and $1.04 for the IFW group. These values assume the outdated infant formula is obtained at no cost.

Case Study 2

Problem. The background for this study is very similar to Case 1 with the exception that the canned liquid product was designed as a geriatric supplement. Its as-fed nutrient composition per liter was 83.7 g protein, 90.9 g fat, 217.3 g carbohydrates, and 2000 calories. Vitamin A, D, and E levels (IU/l) were 5,263, 421, and 47.4, respectively. Calcium, phosphorus, and potassium values (mg/l) were 1,052, 1,052, and 2,456. Fatty acids present in greatest amounts were linoleic, 39.55; oleic 20.47; and caprylic 9.15 g/l. Key ingredients included corn oil and sucrose, making the product likely to induce scouring in neonatal animals. Thus the decision was made to evaluate the product as a nutritional supplement for the growing/finishing pig.

Approach. One hundred sixty pigs were fed either a conventional corn/soybean meal (control) diet or a diet that included 2.8 l/pig/d of the geriatric liquid formula (LF) plus a supplemental dry feed. Pigs were fed from an initial weight of 21.9 kg to a slaughter weight of approximately 117.6 kg. The swine finishing building used in this experiment was a privately owned, modified open-front building. The facility contained 8 pens, each measuring 2.7 m × 7.0 m. Each pen was partially slatted; approximately 35% of the length of each pen was concrete slats (2.4 m), and the remainder was solid concrete (4.6 m). A semi-cylinder, polyvinyl chloride trough 38 cm in diameter × 3.7 m long was placed in four of the pens, over the slatted area, to serve as a feeder for the LF. Each pen was equipped with an automatic nipple waterer and a conventional self-feeder.

Pigs in four of the pens were each fed 1.4 l of LF twice daily after gradually adapting pigs to the product over a period of 2 weeks. In addition, these pigs had access to dry feed from a conventional feeder. The dry feed was formulated to meet or exceed NRC (1979) recommended nutrient levels when fed in combination with the LF. Since most of the nutrients required by the growing pigs were met by consuming the LF, the dry feed contained only corn, supplemental calcium, phosphorus, iron, zinc, and copper. The control diet was a conventional mixture of corn, soybean meal, and the required vitamins and minerals.

Pigs in the LF group initially spent 6 to 8 hours consuming each of their twice-daily allotments of LF. As they gained weight, the LF was consumed in a shorter period of time, until pigs had to compete to receive their respective share. To minimize the competition and to increase the total volume, 0.95 l of water per pig was added with the LF at each feeding time beginning on the 85th day of the experiment.

Diets were fed in two phases: grower and finisher. The grower diets were fed from the start of the experiment until approximately 56.8 kg body weight. The finisher diets were fed from 56.8 kg until slaughter.

Feed for the control pens was automatically delivered to the feeders from a bulk storage bin through a 5-cm auger. To monitor the feed delivered to these pens, a dump scale (Model 600, Kane Manufacturing, Des Moines, IA) was mounted above each self-feeder. Feed was carried to the LF feeders by hand in 22.7 kg bags.

Data from eight pigs were deleted from all statistical analyses for health-related or unsoundness reasons. In addition, no slaughter data were collected from six pigs. From each of the dietary treatments, two pigs were too light to send to market, and one carcass could not be clearly identified at the packing plant.

On the 50th day of the experiment, the feed company incorrectly delivered a load of LF finisher feed into the control bin. The error was not discovered until the pigs were weighed 34 days later, after which the feed was immediately replaced with the correct formulation. Growth performance before the incorrect delivery and after the error was corrected was normal.

Results. Results are presented in table 9.6. Although pigs were weighed every 4 weeks, the data are consolidated into three growth phases (0 to 56 days, 56 to 112 days, and 112 days to slaughter) for ease of interpretation. Pigs fed the LF diet grew faster during the first half of the experiment and weighed more at 84 days; however, growth rates over the entire experiment and slaughter weights were similar for both groups of pigs. Dry feed intake and total dry matter intake were significantly lower in pigs fed the LF diet throughout the experiment. This led to a significant improvement in feed efficiency in pigs fed the LF diet, compared to that of pigs fed the control diet.

Specific carcass data are not presented. Carcass weights were statistically similar between the two groups. Pigs fed the LF diet had less fat, more muscle, a higher percentage of lean cuts, and a higher carcass value.

Implications. Most of the pigs adapted quickly to the LF product and could easily consume 0.75 gallons of LF after a 2-week adaptation period. The three pigs that died and the two that failed to gain weight in the LF group appeared to show no interest in the LF. Refusal by a small number of animals is not unusual when feeding milk-based diets.

Table 9.6 Effect of feeding LF on growth performance in growing and finishing pigs

Item	Control	LF[a]	SE
		Treatment	
Live wt (kg)			
0 day	21.9	21.9	0.29
56 days	58.0	56.4	0.80
112 days	85.7	91.1[h]	1.20
Slaughter[b]	118.2	117.1	1.19
ADG[c] (kg)			
0-56 days	.65	0.62[g]	0.01
57-112 days	0.50	0.62[h]	0.01
112 days-slaughter	0.91	0.72[h]	0.02
0 days-slaughter	0.65	0.65	0.02
Dry ADF[d] (kg)			
0-56 days	1.74	0.34[h]	0.03
57-112 days	2.37	1.43[h]	0.03
112 days-slaughter	2.47	1.40[h]	0.14
0 days-slaughter	2.18	0.77[h]	0.05
LF ADF[e] (l)			
0-56 days	—	2.54	0.13
57-112 days	—	3.02	0.14
112 days-slaughter	—	2.93	0.14
0 days-slaughter	—	2.85	0.13
F/G[f]			
0-56 days	2.71	2.20[h]	0.06
57-112 days	4.79	2.98[h]	0.10
112 days-slaughter	2.65	3.28[h]	0.10
0 days-slaughter	3.29	2.78[h]	0.07

[a]Geriatric liquid formula.
[b]Pigs were slaughtered at 140 or 161 days.
[c]Average daily gain.
[d]Average daily feed intake of dry feed.
[e]Average daily intake of LF.
[f]Feed to Gain ratio; see discussion under Approach.
[g]$p<0.10$.
[h]$p<0.05$

The supplemented dry diet was formulated to meet the nutritional requirements of the growing pig, with the assumption that all pigs would readily consume the assigned amount of LF. This was done to provide a more objective comparison of performance between pigs in the LF and control groups since total nutrient intake for both groups would have been similar. LF pigs consuming only the dry diet, therefore, suffered from low feed intake and multiple nutritional deficiencies.

For Case Studies 1 and 2, the liquid products were delivered in canned form. It was necessary to design and build two can crushers that were powered by the hydraulic system of a farm tractor. (It should be noted that a ready market existed for the compressed cans.) However, concerns that outdated canned products could possibly find their way back into the human food chain has led most companies to offer the products in liquid bulk form only. Alternatively, there is growing interest by several recycling firms in dehydrating the liquids and marketing the resulting dry powders as feedstuffs.

Case Study 3

Problem. In 1997, the FDA issued restrictions on the use of protein material from mammals as a ruminant feedstuff based on the bovine spongiform encephalopathy (BSE) problem in Britain (Federal Register 1997). Excluded from the regulation were pure porcine and pure equine protein, blood and blood products, gelatin, inspected meat products that have been cooked and offered for human food and further heat processed for animal feed use, and milk products. While renderers could expect a decline in sales of certain ingredients such as ruminant-derived meat and bone meal, an increase in the sales of other products or blends of products excluded from the ban might be realized.

Considering this information and the fact that pastured cattle may often experience deficiencies in rumen-undegraded protein intake (Abdalla, Fox, and Seaney 1988; Holden et al. 1994), a trial was designed (Comerford, Harpster, and Baumer 1996) to investigate the value of a blood meal-feather meal blend produced by a large rendering firm in Pennsylvania.

Approach. A complete liquid supplement was formulated to be offered ad libitum to pastured cattle via lick-tank. It contained (as fed) 18% blood/feather meal (80% feather meal, 20% blood meal, 82% crude protein), 42% of a molasses-based premix containing vitamins, NPN, and minerals, 24% water, and 16% vegetable fat. Analysis of the major nutrients in the complete liquid was 57% DM, and on DM basis, 25% CP, 8% NPN, 2.31 Mcal/kg NE_m and 1.82 Mcal/kg NE_g.

In each of 2 years, 65 Holstein steers approximately 22 weeks of age and 180 kg bodyweight were utilized. A representative sample of 5 steers (based on weight and condition) was selected for slaughter to determine initial body composition. The remaining 60 steers were stratified by weight and 20 steers were allotted across weight to each of three treatments: rotationally grazed orchardgrass/alfalfa pastures for 4.5 mo, followed by a high-grain feedlot diet until slaughter (PAST); rotationally grazed orchardgrass/alfalfa pastures for 4.5 mo with ad libidum access to liquid feed, followed by the feedlot diet until slaughter (SUPPL); or fed the high-grain feedlot diet continuously for 263, 307, or 348 days in the first year, and 257, 301, or 342 days in the second year (FEEDLOT). The composition of feedlot diets was set at 15% of the dry matter as corn silage and the remainder as high-moisture shelled corn, soybean meal, and a mineral-vitamin mix containing monensin. Grazing began the first week of May and ended the second week of September each year. Grazing days were 139 in the first year and 133 in the second year (PHASE1). Pastured treatments included three groups per treatment with six, seven, or seven steers per group. The six groups were randomly allocated across available paddocks. Stocking rates were maintained at 0.27 ha per steer during grazing, and the steers were rotated every 5 days. The FEEDLOT steers had ad libidum access to feed in individual feeding gates (American Calan, Northwood, NH).

Following PHASE1, the remaining 45 steers were fed an additional period of 112, 140, or 196 days before slaughter in the first year and 118, 153, or 195 days in the second year. All steers were fed in pens as described for FEEDLOT steers in PHASE1. A 14-day adjustment period was included when PAST and SUPPL treatments began the feedlot phase (PHASE2). Rations were balanced to meet nutrient requirements for large-frame, compensating steers for those in PAST and SUPPL treatments and for large-rame steers in the FEEDLOT treatment. Ration composition was the same for all treatments after 56 days in PHASE2.

Prior to each of the three slaughter dates, all of the steers were stratified by weight within treatment and ultrasonically scanned for fat thickness at the 12th rib. A representative sample of steers within each treatment was selected for slaughter.

Results. Performance results are presented in table 9.7 and partial carcass results in table 9.8.

Across both trial years, growth rate in pastured cattle was increased 33% in steers receiving the liquid supplement. However, liquid supplement intake was highly variable and dependent upon weather conditions and pasture available. The percentage Choice carcasses, fat thickness, dressing percent, yield grade, and final weight were significantly greater for FEEDLOT steers.

Implications. The use of grazing to replace grain in the diets of growing Holstein steers may lower feed costs, but these benefits must be considered with lower slaughter weights and fewer steers reaching Choice

Table 9.7. Weight gain and feed intake of holstein steers in a grazing-finishing trial

Trait	Feeding Phase[1]	Days Fed	Treatment[2]		
			Days PAST	SUPPL	FEEDLOT
ADG	I	136	.52[a]	.69[b]	1.39[c]
(kg/d)	II[3]	56	1.73[a]	1.65[b]	1.14[c]
		112	1.64[a]	1.60[a]	1.17[b]
		140	1.50[a]	1.47[a]	1.12[b]
		196	1.35[a]	1.31[a]	1.04[b]
DMI	I	136			1074.70
(kg)	II	56	609.90[a]	573.10[b]	404.60[c]
		112	1234.70[a]	1172.20[b]	861.30[c]
		140	1520.50[a]	1425.60[b]	1076.50[c]
		196	2107.10[a]	2000.40[a]	1381.90[b]
ADG/DMI	I	136			
	II	56	.16	.16	.15
		112	.15	.16	.14
		140	.14	.15	.14
		196	.13	.13	.13

[a,b,c]Means in the same row and treatment group without a common superscript differ ($p<0.05$).
[1]I = Rotationally grazed for 136 days in grass/legume pastures for PAST and SUPPL treatments and high concentrate rations fed in a feedlot for 136 days for the FEEDLOT treatment.
[2]PAST = Rotationally grazed on grass/legume pastures for 136 days with N = 40 in phase I and n = 30 in phase 2; SUPPL = Rotationally grazed on grass/legume pastures for 136 days with ad libitum access to molasses-based liquid supplements with N = 39 for phase I and 29 for phase II; FEEDLOT = fed high concentrate diets in a feedlot with N = 40 for phase I and N = 30 for phase II.
[3]Steers were serially slaughtered in phase II with N = 89 for 56 and 112 days, N = 59 for 140 days and N = 29 for 196 days.

grade to fully determine its profitability. Liquid supplements provided ad libitum to lightweight grazing Holstein steers will significantly increase weight gains, but the inconsistency of intake and overconsumption are problems requiring further research. It did not appear that the Holstein steers effectively used grazing alone for weight gain, and supplemental protein feeding was required to reach adequate levels of production. The animal-derived by-product proteins used in this study were effective in promoting additional growth performance in the grazing steers.

Case Study 4

Problem. A local cannery operation must decrease the practice of food-waste disposal by land-spreading. Significant steps taken by the company

Table 9.8. Carcass traits in holstein steers in a grazing-finishing trial

Treatment[1]	Slaughter group[2]	Dressing percent (%)	Ribeye area (cm²)	Fat thickness (cm)	Percent Choice
PAST	$\overline{X}$	58.2	63.9	.48	60.0
	1	57.3	63.7	.33	30.0
	2	58.4	62.6	.51	70.0
	3	58.9	65.5	.60	80.0
SUPPL	$\overline{X}$	58.4	68.2	.51	51.2
	1	56.7	65.1	.46	30.0
	2	58.9	69.9	.55	80.0
	3	59.6	69.7	.51	43.6
FEEDLOT	$\overline{X}$	59.9	69.1	.75	80.0
	1	59.3	69.7	.68	60.0
	2	59.8	67.2	.80	90.0
	3	60.5	70.6	.75	90.0

[1]See table 9.7.
[2] 1 = 112 days postgrazing in the feedlot for PAST (N = 10) and
SUPPL (N = 9), and 247 days in the feedlot for FEEDLOT (N = 10);
2 = 140 days postgrazing in the feedlot for PAST (N = 10) and
SUPPL (N = 10), and 276 days in the feedlot for FEEDLOT (N = 10);
3 = 196 days postgrazing in the feedlot for PAST (N = 10) and
SUPPL (N = 10), and 332 days in the feedlot for FEEDLOT (N = 10).

to increase the acceptability of the waste stream as a potential animal feed included elimination of hardware contamination and a combination grinding and filter press installation. This resulted in a more dense and uniform product that averaged approximately 20% dry matter versus the pretreatment level of 10%.

Approach. A trial was initiated (Harpster 1997) to compare silage systems for feedlot cattle. A significant portion of the plant output included the residues of carrot and potato processing. Given the seasonal nature of these crops, ensiling was chosen as the management system. The pressed food waste, which contained approximately 90% in vitro digestible dry matter, was combined either with chopped, low quality hay that had been stored in round-bale form, or corn fodder chopped from a field where grain had been harvested.

Waste-forage silages were prepared and placed in Ag Bags®. The proportions of food waste and forage were based on a desired final silage dry matter content of 32 to 35%.

Food waste was received, mixed with forage, and ensiled as generated by the plant and a 2-week period was required to obtain sufficient quantities. Fiber sources and food waste were combined in 2- to 4-ton batches in a horizontal feed mixer, thoroughly mixed, and immediately ensiled in a Silo-Press® silage bag. Approximately 47 tons of hay-based silage and 56 tons of cornstalk-based silage were prepared. Pre-ensiling pH values were generally low, usually below 5.0. Keeping quality of all three silages was acceptable. Mean pH values from weekly samples collected during the feeding trial were corn silage, 3.89; food waste/hay silage (FW/H), 4.58; and food waste/cornstalk silage (FW/CS), 4.78.

In a feeding trial with crossbred steers, 10 calves were assigned to each of three treatments: corn silage control (C), FW/H, and FW/CS. Initially, each silage was fed at approximately 60% of the total ration dry matter; the remainder being cracked corn, soybean meal, and a mineral/vitamin mix. Preliminary observations of feeding behavior indicated excellent intake of all diets, although steers assigned to the cornstalk-based silages exhibited some sorting of the ingredients of the ration. This problem was related to the large particle size of the stalks as a result of the field chopping procedure. Later in the feeding period during the fattening phase, the proportions were reversed (40% silage:60% concentrates). The protein and macro-mineral content of the primary feedstuffs included in this evaluation are in table 9.9.

Steers were individually fed one of the three treatments using Calan® electronic feeding doors. Feed offered was weighed daily. Weighbacks and feed samples were taken weekly, dried, and ground. Samples were composited on a period (28 days) basis. All samples were analyzed for dry matter, crude protein, and ADF. Steers were weighed every 28 days and slaughtered in three groups with three steers per treatment slaughtered at 168, 196, and 224 days on feed. Ultrasound scanning was utilized to select the three fattest steers in each treatment group at each slaughter time.

Results. Cumulative performance data are presented in table 9.10 for the eight feeding periods representing 224 days on feed. Period 1 per-

Table 9.9. Crude protein and mineral content of by-product trial feeds

| Feedstuff | Analyses, % of Dry Matter | | | | |
	Crude Protein	Calcium	Phosphorus	Potassium	Magnesium
Corn silage	8.0	0.31	0.27	1.22	0.22
Shelled corn	10.0	0.03	0.31	0.33	0.11
Soybean meal	54.0	0.29	0.71	2.36	0.33
FW/H silage[1]	8.9	0.47	0.21	1.58	0.15
FW/CS silage[2]	6.2	0.31	0.14	0.93	0.12

[1]Food waste/hay silage.
[2]Food waste/corn stalk silage.

formance data indicate that cattle adapted quickly to the by-product silage rations with gains actually exceeding those of cattle fed the corn silage control. At 168 days on feed, gains relative to the control were 82.1% for the food waste/hay silages and 76.1% for the food waste/stalk group. At 196 days, gains relative to control were 88.96% for the food waste/hay silages and 82.80% for the food waste/stalk group. At 224 days, gains relative to control were 91.84% for the food waste/hay silages and 83.33% for the food waste/stalk group.

Carcass data were collected for the three slaughter groups. There were no differences in ribeye area, lean firmness, lean color, or percentage kidney, heart, and pelvic fat. Carcasses from steers fed the control diet were heavier, fatter, had higher yield grades, and slightly higher

Table 9.10. Daily gain of steers fed control and by-product silages

Period	Treatment[a]	Steers	Days	Cumulative Gain
1	1	10	28	3.45
	2	10	28	3.69
	3	10	28	3.56
2	1	10	56	4.12
	2	10	56	4.07
	3	10	56	3.34
3	1	10	84	4.34
	2	10	84	3.88
	3	10	84	3.50
4	1	10	112	4.22
	2	10	112	3.59
	3	10	112	3.21
5	1	10	140	4.19
	2	10	140	3.49
	3	10	140	2.94
6	1	10	168	3.97
	2	10	168	3.26
	3	10	168	3.20
7	1	7	196	3.63
	2	7	196	3.23
	3	7	196	3.01
8	1	4	224	3.50
	2	4	224	3.20
	3	4	224	2.92

[a]1 = corn silage control; 2 = hay/food waste silage;
3 = stalk/food waste silage.

USDA quality grades. Fat color scores were somewhat higher in carcasses from steers fed the by-product silage diets. The carrot waste may have been responsible for a higher incidence of "casty" or yellowish-colored fat in some animals consuming the by-product diets. On average, however, this was not of sufficient magnitude to create a marketing problem.

Implications. By-product silages containing food industry waste can be a viable alternative to corn silage as a feedstuff for ruminant animals. The time (2 weeks) to obtain enough food waste from the single plant source to prepare 100 tons of silage was problematic. It may be possible to stockpile the food waste in a deep pit before beginning the ensiling process to make more efficient use of labor and machinery and shorten the time required to fill a given silo structure. Although the nutritive value of the waste silages was approximately 20% less than corn silage, production costs were lower. In addition, the major factor limiting the energy value of the waste silages is the low dry matter content of the food waste (approximately 15-20%). This limits the inclusion level of the food waste since, when combined with roughage, the total mixture must contain approximately 30 to 35% dry matter for proper anaerobic fermentation. Thus, if cost-effective technology could be found to more completely de-water the food waste at the plant of origin, a larger percentage of this high-energy material could be blended with the roughage source. Depending on the energy level of the roughage base used it should be possible to create silages at least equivalent to corn silage in nutritive value.

Future Needs

The potential to produce animal products from food-processing residuals is enormous and only partially realized at this time. While a lack of sound nutritional information is responsible for a portion of this underutilization, other factors are involved. For example, in this electronic age it would be extremely useful to establish list-serves on the Internet where food industry suppliers could advertise the type, quantity, seasonality, and location of materials they have available to livestock producers. There is also a continuing need to work directly with food processors to provide at least a rudimentary understanding of the nutritional needs of farm animals and the changes needed in waste stream management to facilitate feed by-product use.

Another high priority is to establish more uniform feed control laws from state to state as they relate to food-processing residuals. The laws of many states are more restrictive than federal guidelines. For an excellent discussion of this issue as well as a summary of state requirements, readers are referred to Polanski (1996).

Several more direct nutritional needs can be identified. With the constant feed testing required of highly variable food-processing waste, it would be desirable to make increasing use of near infrared reflectance technology, especially on wet, as received, samples. There is presently a lack of adequate calibration sample sets of many by-products to allow meaningful use of the technology. In a similar vein, more complete analyses of nonconventional feeds are needed to allow more sophisticated ration balancing. Databases that include those fractions required by the Cornell Net Protein/Carbohydrate System of feed evaluation would be extremely useful (Hinders 1995).

Finally, there is a need to address the question of how much processing one can justify in the treatment and handling of a variety of food-processing wastes. Drying, grinding, pelleting, extruding, ensiling, etc. are processes that can enhance feed value but at what cost/benefit ratio?

Conclusion

There is considerable promise in increasing the use of industrial by-products as animal feeds. Food-processing industries are more willing to cooperate with potential users given current environmental pressures. Problems remain, however, in obtaining reliable and consistent sources of by-products. Nutritional management must be at a high level to cope with the variability encountered. High moisture wastes are excellent candidates for ensiling when combined with dry roughages. Further research is needed in designing practical feeding systems and required processing techniques. As always, comparative feed economics will determine the degree of by-product in the future.

Acknowledgements

The author gratefully acknowledges the support of the Department of Dairy and Animal Science administration and the livestock faculty, staff, and graduate students of Pennsylvania State University.

References

Abdalla, H. O., D. G. Fox, and R. R. Seaney. 1988. Variation in protein and fiber fractions in pasture during the grazing season. *J. Anim. Sci.* 66:2663.

Bath, D., J. Dunbar, J. King, S. Berry, and S. Olbrich. 1997. Byproducts and unusual feedstuffs. *In:* Feedstuffs Ref. Issue 69(30):32. Minnetonka, MN: Miller Publishing Co.

Brandt, R. C. and K. S. Martin. 1994. The Food Processing Residual Management Manual. PA Dept. of Env. Resources Publ. No. 2500-BK-DER-1649.

Comerford, J. W., H. W. Harpster, and V. I. Baumer. 1996. Effects of grazing and protein supplementation for Holstein steers. *J. Anim. Sci.* 74(Suppl 1): 254.

Eastridge, M. L. 1995a. Economical value of alternative feeds based on nutritive composition. *In:* Proc. of Second National Alternative Feeds Symposium. Sept 24-26, St. Louis, MO.

Eastridge, M. L. (ed.). 1995b. Proc. of Second National Alternative Feeds Symposium USDA and Univ. of Missouri-Columbia. Sept. 24-26, St. Louis, MO.

Federal Register. 1997. Regulation 62 FR 30936. June 5, 1997.

Grasser, L. A., J. G. Fadel, I. Garnett, and E. J. DePeters. 1995. Quantity and economic importance of nine selected by-products used in California dairy rations. *J. Dairy Sci.* 78:962-971.

Harpster, H.W. 1997. Unusual silages based on food processing by-products. *In:* Proc. Silage to Feedbunk-North American Conference, Feb. 11-13, Hershey, PA. NRAES Publ. 99.

Hinders, R. 1995. Cornell system useful in evaluation of rations containing by-products. *Feedstuffs* 67(47):12.

Holden, L. A., L. D. Muller, G. A. Varga, and S. L. Fales. 1994. Estimation of intake in high producing Holstein cows grazing grass pasture. *J. Dairy Sci.* 77:2332.

NRC. 1979. Nutrient Requirements of Swine. 8th Rev. Ed. National Academy Press, Washington, D.C.

Polanski, J. 1996. Legalizing the feeding of nonmeat food wastes to livestock. *In:* Proc. of Food Waste Recycling Symposium. M. Westendorf and E. Zirkle, (eds). USDA, Rutgers Univ., and New Jersey Dept. of Ag. Jan. 22,23. Atlantic City, NJ. pp.3-9.

Rogers, G. M. and M. H. Poore. 1994. Alternative feeds for reducing beef cow feed costs. *Vet. Med.* 89(11):1073-1084.

Swope, R. L., H. W. Harpster, R. S. Kensinger, and V. H. Baumer. 1995. Nutritive value of human infant formulas for young calves. *J. Anim. Sci.* 73(Suppl 1): 249 (Abstr.).

Westendorf, M. and E. Zirkle (ed.). 1996. Proc. of Food Waste Recycling Symposium. USDA, Rutgers Univ. and New Jersey Dept. of Ag. Jan. 22,23. Jan. 22-23, Atlantic City, NJ.

Wilson, L. L. and P. G. Lemieux. 1980. Factory canning and food processing wastes as feedstuffs and fertilizers. *In:* M. Bewick (ed.) Handbook of Organic Waste Conversion. pp. 243-267. New York: VanNostrand Reinhold Co.

10

Sweetpotatoes and Associated By-products as Feeds for Beef Cattle

by Matthew H. Poore, Glenn M. Rogers,
Barbara L. Ferko-Cotten, and
Jonathan R. Schultheis

Introduction

Utilizing alternative feed sources allows many beef producers to reduce production costs (Rogers and Poore 1994). This chapter centers on the nutritional and health aspects of utilizing cull sweetpotatoes and their related by-products as alternative feedstuffs for beef cattle.

Sweetpotatoes (*Ipomoea batatas* Poir.) are an important food crop ranking seventh among crops grown for food worldwide (FAO 1986). Sweetpotatoes are grown extensively in the southern United States (figure 10.1), with approximately 1.35 billion pounds produced in 1997 (NCDA 1998).

To avoid confusion, the one-word spelling of "sweetpotato" first was established by the American Society for Horticultural Science (ASHS 1991) because while sweetpotatoes are not actually potatoes, the two-word spelling can be interpreted as an adjective modifying a noun (i.e., a sweet-tasting potato). Sweetpotatoes are storage roots that have different storage and growth requirements from potatoes, which are modified stems or tubers, and these crops should not be confused by shippers, distributors, warehouse workers, or consumers. Sweetpotatoes are a warm-season crop, growing best between 21° and 29°C (Peirce 1987), and are typically produced in the southern states. The top five sweet-potato-producing states, which account for nearly 90% of the U.S. sweet-potato crop, are North Carolina, Louisiana, California, Mississippi, and Texas (NCDA 1996).

The terms "yam" and "sweetpotato" are often utilized interchangeably in the United States; however, these plants are representatives of two distinct species. Sweetpotatoes are smooth storage roots with thin

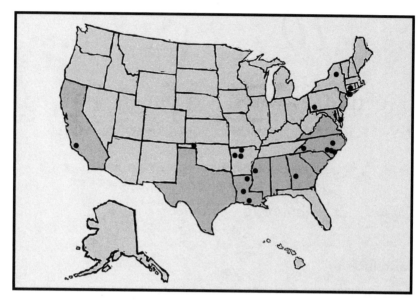

Figure 10.1. Sweetpotato producers and processing plants located in the United States. The top 10 producers in order and production in millions of pounds (1994 data): North Carolina, 527; Louisiana, 304; California, 168; Mississippi, 93.5; Texas, 83.7; Alabama, 79.8; Georgia, 36; South Carolina, 21.9; New Jersey, 15.4; Virginia, 8. Note: • = location of processing plants.

skin, tapered ends, and a moist, sweet taste. Yams (*Dioscorea sp.*), however, are long, rough, scaly tubers that have a more characteristic dry and starchy taste. Yams require tropical growing conditions and must be imported, while sweetpotatoes can be grown in both temperate and tropical climates (Wilson and Collins 1993).

Sweetpotatoes are utilized for baking and are processed for freezing, chips, baby food, and canning. Twenty processors are located on the west and east coasts; however, the majority of processors are near areas of sweetpotato production, particularly where the product is canned (figure 10.1).

Sweetpotato-processing waste takes various forms such as peel waste, cull sweetpotatoes, screen waste, and several others (table 10.1). Sweetpotato-processing waste provides an excellent source of readily digestible carbohydrates for cattle producers located in close proximity to the processing plants; however, high moisture content greatly limits the distance that the residues can be economically hauled. Sweetpotato-processing waste is generally received from processing plants as a liquid or slurry mixture and stored in open pits, allowing for a constant supply for cattle diets after a natural fermentation that readily occurs (Rogers and Poore 1997).

Table 10.1. Sweetpotato by-products available for animal feeding

PRODUCT	SOURCE
Sweetpotato By-products [1]	
Whole/chopped cull sweetpotatoes (Morrison 1946; Seath et al. 1947)	Unmarketable sweetpotatoes due to insect damage, bruising, etc., may be sliced or chopped
Sweetpotato peels [2] (Sistrunk and Karim 1977)	Peels removed by boiling in sodium hydroxide solution (lye); digested peel removed by rubber scrubbers w/o water
Sweetpotato cannery waste (SPCW) (Woolfe 1992; Rogers et al. 1997)	Peels and chunks of sweetpotato resulting from steam-peeling (without alkali), and then fermented in open pits to pH 3-4
Sweetpotato vines (Seath et. al 1947)	Cows graze vines as pasture
Sweetpotato silage (Ruiz 1982)	Stems, leaves, roots stacked and fermented as silage
Stillage waste (Woolfe 1992)	Nonvolatile fraction of material used for distillation of spirits or industrial alcohol
Sweetpotato chips (dehydrated) (Wu 1980)	Sun-cured, fine ground chips of sweetpotato
Sweetpotato cannery solids (Rogers and Poore 1997)	Steamed whole sweetpotatoes and large chunks of sweetpotatoes from canning process, along with some peels

[1] Due to large expenditures for drying, current interest focuses on utilizing wet wastes
[2] Most processing plants currently utilize steam-peeling rather than the addition of lye.

Cull sweetpotatoes (figure 10.2a and 10.2b) are those that fail to meet minimum requirements for tablestock, seedstock, or processing as a result of bruising, disease, or insect damage, and are an important by-product of the industry. Usually, disposal of cull sweetpotatoes does not present a problem; however, in years when large amounts of sweetpotatoes are produced, disposal is both expensive and difficult. Large quantities of cull sweetpotatoes are disposed of in three major ways: feeding to livestock (figure 10.3), incorporation into landfills, or spreading on cropland. Dumping of sweetpotatoes near streams, marshes, wells, and gravel pits may contaminate surface water and ground water sources, and decomposing waste sweetpotatoes also emit objectionable odors and provide a breeding ground for many insects, especially small flies. Feeding cull sweetpotatoes to livestock is an ideal disposal method because it may provide an economic return for only a minimum of cost. Cull sweetpotatoes and sweetpotato by-products are currently being utilized in livestock rations in the southeastern United States and in many areas of Latin America (Espinola 1992).

a

b

Figure 10.2. Cull sweetpotatoes (a) and moldy culls (b) similar to those causing atypical interstitial pneumonia.

Figure 10.3. Cows eating cull sweetpotatoes from a pile with access limited by an electric fence.

North Carolina is the nation's leading producer of sweetpotatoes, with more than 31,000 acres harvested in 1997, yielding 496 million lbs. Approximately 40% of these sweetpotatoes undergo a steam-peeling cannery process, which produces more than 64 million lbs of sweetpotato cannery waste (SPCW, figure 10.4a). Sweetpotato cannery waste has an estimated energy value equivalent to corn and has been recycled through cattle production systems located near the processing plants. Thus, it provides an inexpensive energy source for cattle while reducing the costs and limiting the negative environmental impacts and costs associated with SPCW disposal in landfills or wastewater treatment facilities (Rogers and Poore 1997).

Unmarketable (cull) sweetpotatoes can also be utilized as livestock feed, thus increasing the per acre return from U.S. sweetpotatoes grown primarily for human consumption (Briggs et al. 1947). Kummer (1943) describes an easily constructed mechanical device for slicing and shredding large volumes of cull sweetpotatoes, allowing for rapid drying in favorable weather conditions. Various other sweetpotato by-products that have been utilized as livestock feeds are dehydrated sweetpotato meal, sweetpotato chips (flakes), and sweetpotato foliage offered fresh or as silage (Espinola 1992; Wu 1980; Singletary et al. 1950).

a

b

Figure 10.4. Unloading sweetpotato cannery waste into a storage pit (a) and cows eating sweetpotato cannery waste free-choice (b).

Feeding Value

Sweetpotatoes contain large amounts of starch and sugars and are used mainly as energy supplements in livestock feeds. The commonly published average dry matter content of sweetpotatoes is 31% (Bath et al. 1998), while the USDA sweetpotato database indicates a value of 27% (USDA 1996). However, the Beauregard variety, which is widely grown in North Carolina and Louisiana, only contains 18 to 24% dry matter (Walter 1997; Dukes et al. 1987; Rogers and Poore 1997). Generally, sweetpotato varieties grown for human consumption in the United States have lower dry matter content than those that are used for ethanol production (Rolston et al. 1987). Sweetpotatoes are also rich sources of vitamins A and C, thiamin, riboflavin, niacin, and carotene (Dominguez 1991); however, they are low in protein, calcium, and phosphorus. Nutrient composition of sweetpotato by-products varies greatly depending upon the portion of the plant (vines or roots) and the processing method utilized. For example, steamed sweetpotato solids had an average crude protein value of 7.3%, while SPCW slurry from the same plant had a higher average crude protein value of 10.7% (Rogers and Poore 1997). Dried sweetpotato tops have an average crude protein value of 13.9% (Bath et al. 1998). Table 10.2 shows nutrient composition for selected sweetpotato by-products.

Briggs et al. (1947) found that dried sweetpotatoes were highly palatable to steers; no refusals occurred when included in the ration. Based on total digestible nutrient content, dried sweetpotatoes had 92.3% the value of No. 3 corn. Similar results from a 1950 study by Singletary et al. (1950) showed that high-quality dehydrated sweetpotato meal used as the carbohydrate feed in a balanced ration was worth 90 to 95% as much as corn for finishing calves. No significant differences in ADG or dressing percentages were observed between rations that included dehydrated sweetpotatoes as one-third, two-thirds, or all of the carbohydrate ration.

Results from an 84- day feeding trial utilizing a free-choice mixture of 90% SPCW plus 10% broiler litter along with free-choice ryegrass hay, showed comparable average daily gains for the broiler litter plus SPCW ration and the control ration (7 lbs/hd/day corn plus soybean meal and free-choice ryegrass hay). Once final live weights were adjusted to a constant dressing percentage (carcass weight/0.50), the SPCW plus litter treatment resulted in the highest rate of gain (table 10.3; Rogers et al. 1999).

In a study with stocker heifers (initial weight 500 lbs) fed cottonseed hull-based total mixed rations, utilizing 15% (dry matter basis) chopped

Table 10.2. Nutrient composition of selected sweetpotato products

| | Sweetpotatoes | | | |
	Whole	Cannery Solids	Tops, Dried	SPCW
Dry Matter (%)	27.16[a]	17.60[c]	91.0[b]	8.41[c]
Crude Protein (%)	6.08[a]	7.31[c]	13.9[b]	10.66[c]
Acid Detergent Fiber (%)	8.00[b]	NA	26.0[b]	12.98[c]
TDN (%)	80.00[b]	NA	57.0[b]	NA
Net Energy, Maintenance (Mcal/lbs.)	0.85[b]	NA	0.55[b]	NA
Net Energy, Gain (Mcal/lbs.)	0.57[b]	NA	0.25[b]	NA
Calcium (%)	0.08[a]	0.21[c]	NA	0.32[c]
Phosphorus (%)	0.10[a]	0.16[c]	NA	0.26[c]
Sodium (%)	0.05[a]	0.07[c]	NA	0.08[c]
Magnesium (%)	0.04[a]	0.14[c]	NA	0.14[c]
Sulfur (%)	0.13[d]	0.10[c]	NA	0.16[c]
Potassium (%)	0.75[a]	1.69[c]	1.0[d]	3.04[c]
Copper (ppm)	6.20[a]	5.00[c]	NA	10.00[c]
Iron (ppm)	21.72[a]	138.00[c]	NA	803.50[c]
Manganese (ppm)	13.07[a]	9.00[c]	NA	35.50[c]
Zinc (ppm)	10.31[a]	17.00[c]	NA	22.00[c]
Ash (%)	3.50[a]	NA	11.3[b]	7.92[c]

[a] Values taken from USDA 1996.
[c] Values taken from Rogers and Poore 1997.
[b] Values taken from Bath et al. 1998.
[d] Values taken from Dunbar et al. 1990.
NA- Not Available

Note: These values reflect an average of sweetpotato varieties. Different varieties of sweetpotato differ in dry matter content. The Beauregard variety, which is widely grown in North Carolina, contains approximately 4 to 8% less dry matter than the values indicated above (Walter 1997).

Table 10.3. Performance of Holstein steers fed a control ration, free-choice sweetpotato cannery waste supplemented with soybean meal or sweetpotato cannery waste neutralized with 10% broiler litter (as is basis). Taken from Rogers et al. 1999.

Item	Control	SPCW	SPCW + Litter
Adjusted average daily gain (lb/d)	1.94[a]	1.58[b]	2.24[c]
Dry Matter Intake (lb/day)	20.40[a]	15.60[b]	24.40[c]
Feed/Gain	9.22	8.31	9.68
Sweetpotato Cannery Waste Intake (lb/day, wet basis)	—	84.00	101.60

[a,b,c] $p < 0.05$

whole sweetpotatoes (fermented in large plastic containers for at least 2 weeks) as a partial substitute for corn resulted in similar average daily gain (2.08 v. 2.07 lbs/day for control and sweetpotato rations, respectively), and improved feed efficiency (10.22 v. 9.08 lbs feed/lbs gain for the control and sweetpotato rations, respectively) as compared to the control containing 47% ground corn (Poore et al. 1998).

Sweetpotatoes and their associated by-products are a good source of readily degradable carbohydrate (energy), but are marginal sources for protein and minerals. Utilization is usually limited by sweetpotatoes' high moisture content. Dried sweetpotatoes may be used as a high proportion of the carbohydrate in diets, but wet products will generally be limited to 15 to 25% of diet dry matter. Studies evaluating wet sweetpotato by-products as a portion of a TMR show they are successfully substituted for corn with comparable animal performance. Feeding free-choice cannery slurry is possible if the acidic material is neutralized with broiler litter or perhaps by using some other neutralizing agent.

Identified Health Concerns

Respiratory System

The toxic effects of moldy or rotten sweetpotatoes (figure 10.2) on the bovine respiratory system have been recognized for more than 40 years in both the United States and Japan (Wilson et al. 1970). Cattle may show symptoms of severe respiratory distress, rapid respiratory and heart rates, and the presence of frothy exudate around the mouth as soon as 1 day after feeding on affected sweetpotatoes (Peckham et al. 1972). The characteristic clinical and pathologic findings are indistinguishable from acute bovine pulmonary emphysema and edema (ABPEE) or "fog fever," and include severe alveolar edema, hyaline membrane formation, and interstitial emphysema. The disease was characterized as pulmonary adenomatosis and currently has been reclassified as atypical interstitial pneumonia. Significant herd losses due to this respiratory disease have been reported since at least 1928 (Peckham et al. 1972). Although atropine, steroids, diuretics, diethylcarbamazine, antibiotics, and vitamins have been used in the treatment of atypical interstitial pneumonia, no specific treatment significantly alters the outcome of the disease. Until safer methods for feeding are developed, the best management practice is to avoid feeding moldy or decaying sweetpotatoes to livestock (Wilson et al. 1981).

Prior to World War II, Japanese investigators isolated four substituted furans from sweetpotatoes infected with the black rot fungus *Ceratocystis fimbriata*: ipomeamarone, ipomeanine, furoic acid, and batatic acid. Ipomeamarone, the only metabolite found to be toxic, was shown to be hepatotoxic for mice. Fungal infection, contact with certain chemicals (mercuric chloride and iodoacetates), and insect damage stimulate the sweetpotato to synthesize ipomeamarone (Wilson 1973).

An outbreak of lung disease that proved fatal for 69 of 275 Hereford cattle in Tifton, Georgia (following ingestion of cull sweetpotatoes) again prompted further research on sweetpotato toxins (Wilson 1973). Fungal isolates from several genera were isolated from the moldy tubers: *Botryodiplodia, Fusarium, Aspergillus, Penicillium, Rhizopus,* and *Mucor.* The causal relationship was determined after sweetpotatoes infected with an isolate of *Fusarium solani* were fed to the herd and identical disease symptoms were observed (Peckham et al. 1972). Sweetpotatoes inoculated with *Fusarium solani* produced several abnormal metabolites including ipomeamarone, ipomeamaronol, and previously unknown lung edema toxins known collectively as the "lung edema factor" (Wilson 1973).

Sweetpotatoes produce several stress metabolites known as phytoalexins in response to exogenous stimuli such as mechanical injury, chemical irritation, and nematode or fungal infection. In sweetpotatoes infected with *F. solani* or *F. oxysporum,* these phytoalexins are metabolized to produce four closely related compounds known as the "lung edema factor," which is associated with severe edema and proliferative alveolitis of the lungs. Of these compounds, 4-ipomeanol is thought to be the most active in producing lung toxicity, with cattle being particularly susceptible (Hill and Wright 1992). Experimentally, synthetic 4-ipomeanol has produced clinical disease and pulmonary lesions identical to those produced by purified ether extracts of moldy sweetpotatoes. Likewise, cattle given intraruminal administration of 4-ipomeanol developed a respiratory syndrome clinically and histologically indistinguishable from atypical interstitial pneumonia. Heifers receiving lesser amounts of 4-ipomeanol experienced fewer pulmonary changes. Such studies indicate that 4-ipomeanol is the major toxic principle in moldy sweetpotatoes responsible for development of acute interstitial pneumonia (Doster et al. 1978). While it is known that 4-ipomeanol occurs in moldy sweetpotatoes, it is not known when the toxin will occur at levels sufficient to cause problems. Sometimes, sweetpotatoes with characteristic *F. solani* lesions have potentially problematic levels (>100 ppm) of 4-ipomeanol, while at other times similar appearing sweetpotatoes have low or undetectable levels (M.H. Poore, unpublished data). Work is currently underway to better predict when toxin levels in packing house culls are high enough to cause problems, and in what type of tissue the toxin is likely to occur. Processing effects on 4-ipomeanol levels in cull sweetpotatoes also are currently under investigation.

Choke

Choke occurs commonly when feeding whole cull sweetpotatoes to cattle, as turgid objects can easily slip from an animal's teeth and lodge in the throat. Allowing cull sweetpotatoes to wilt, soften, or ferment, or chopping them before feeding will help to reduce the incidence of choke. Cull sweetpotatoes may likewise lodge within the esophagus, creating a physical obstruction to eructation and thus produce secondary ruminal tympany or free-gas bloat (Radostits et al. 1994).

Laminitis

Sweetpotatoes and their by-products are rich sources of rapidly fermentable nonstructural carbohydrates and may cause problems with laminitis if not gradually introduced into the ration. Feeder cattle and heifers around the time of parturition are especially susceptible to rapid dietary changes and the development of laminitis. To aid in the prevention of laminitis, it is advisable to gradually incorporate sweetpotatoes or their by-products into the ration, especially for pregnant heifers and young bulls (Radostits et al. 1994). Adequate amounts of forage should also be readily available during or following feeding to help in prevention of acidosis (Greenough et al. 1981). Limiting the rate of dietary inclusion of sweetpotatoes and their by-products can also help in preventing acidosis and laminitis. (See sample rations under Recommendations.) Proper composition of the ration and good feeding and bunk management are likewise essential components in any feeding program.

Dental Erosion

Dental erosion of dietary origin is rare in cattle (Rogers and Poole 1987); however, studies have shown that use of sweetpotato cannery waste free-choice (figure 10.4b) is directly associated with the development of severely eroded and blackened incisors (figure 10.5a and 10.5b; Rogers and Poore 1997). This problem was originally identified in one large cow/calf operation feeding SPCW free-choice as a majority of the diet about 6 months of the year. This operation had been experiencing poor performance especially among yearling heifers that had very low body condition and poor breeding rates. A thorough examination revealed that about two-thirds of the cows exhibited moderate to severe dental erosion. Several risk factors were investigated, but SPCW was later confirmed as the cause when another herd also feeding SPCW without the other risk factors was investigated, and identical dental erosion was observed. Figure 10.6 shows radiographs from a normal aged cow with worn teeth, and from a cow being fed SPCW. Note that the cow fed SPCW had open communication with the interior of the tooth, and that apical abscesses are visible in the mandible.

The acidic nature (pH 3.2 to 3.4) and the high lactic acid content of SPCW (up to 2.28% on an as fed basis; Rogers et al. 1997) are the major

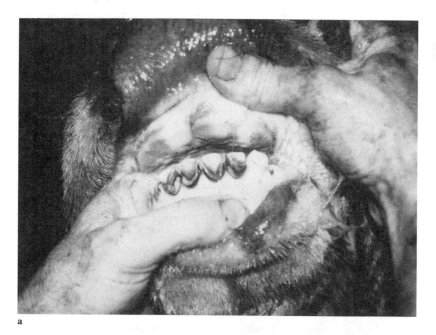

a

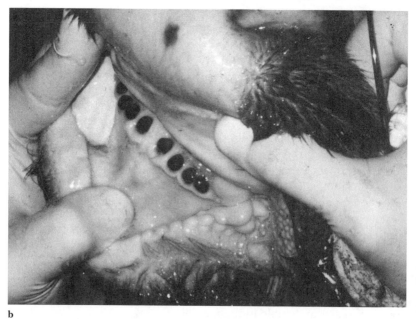

b

Figure 10.5. Young (a) and middle-age (b) cows with severe enamel erosion as a result of feeding sweetpotato cannery waste (Rogers and Poore 1997).

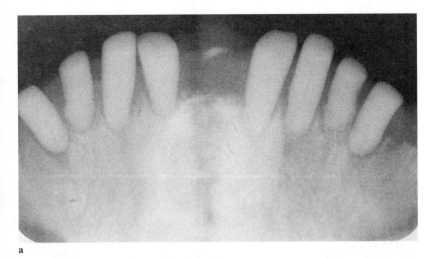

a

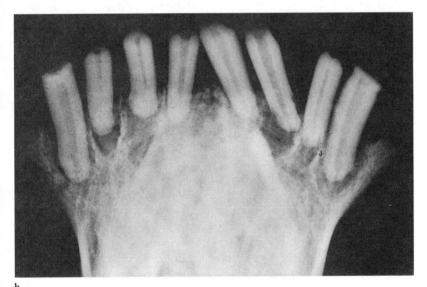

b

Figure 10.6. Radiographs of a cow with normal incisor wear (a), and a cow with incisors eroded as a result of feeding free-choice sweetpotato cannery waste (b).

contributors to its erosive potential. In vitro studies measured enamel erosion as grams of calcium ion removed from a specified area (3 mm diameter disc) of the enamel surface of incisor teeth upon exposure to either lactic acid or SPCW for a short period of time (figure 10.7). The figure shows the amount of calcium released during a 30- or 60-second

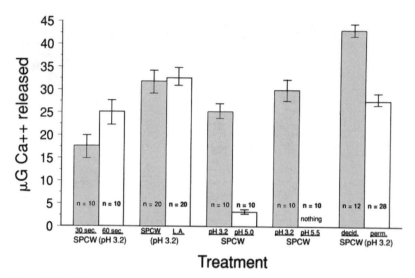

Figure 10.7. Effects of sweetpotato cannery waste and lactic acid on calcium loss from bovine teeth (Rogers et al. 1997).

etch in the first bars, and all others show a 60-second etching time. Nearly identical amounts of calcium were removed by the lactic acid and SPCW solutions (both pH 3.2), with calcium removal 56% higher in deciduous than in permanent teeth. Neutralization of SPCW to pH 5.5 (in vitro) totally eliminated calcium removal (Rogers et al. 1997).

A feeding trial with Holstein steers (Rogers et al. 1999) compared feeding SPCW as a slurry (pH 3.2) free-choice to a 90% SPCW/10% broiler litter slurry (pH 4.0), and a control of corn plus soybean meal and hay. Enamel erosion and staining was only slight for the SPCW/broiler litter slurry compared to the control ration after 84 days. In contrast, SPCW (pH 3.2) produced enamel erosion and discoloration after only 28 days in most animals, and by 84 days, erosion was severe (figure 10.8).

Acid erosion of bovine teeth may be a problem when other highly acidic vegetable-processing slurries are fed free-choice because of their high lactic acid content (Sauter, et al. 1985). As pH of etching solution was varied between 3.2 and 4.75 there was a large decrease in Ca^{2+} release to the etching solution (figure 10.9). Between pH 3.2 and 4.0 there was a 70% decrease in the etch rate, supporting our findings and suggesting that neutralizing vegetable-processing slurries to above pH 4.0 before feeding free-choice is advisable.

Representative Steer from Each Treatment

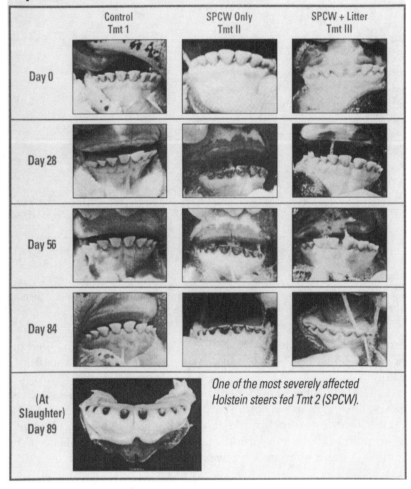

Figure 10.8. Teeth of representative calves fed a control ration, sweetpotato cannery waste or 90% sweetpotato cannery waste/10% broiler litter.

Other Digestive Problems

Sweetpotato by-products processed using sodium hydroxide (caustic lye), are often delivered to the site of feeding at pH values between 11 and 12. Standing in a pit for 5 or more days can cause pH values to decline below 4.0. As a result of feeding relatively fresh material, cattle may experience a "nutritional roller coaster" and subsequent digestive

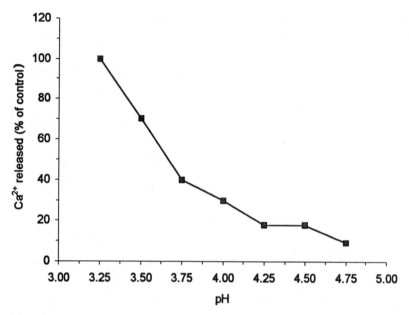

Figure 10.9. Amount of calcium ion released from bovine teeth in response to etching with lactic acid solutions of varying pH. Values are a percentage of loss at pH 3.2 (Rogers et al. 1999).

problems, due to extreme pH variation (Mehren 1996). Fermented by-products should be allowed to stand in a pit and ferment to a stable pH before feeding.

Unwanted artifacts and foreign objects such as ball bearings, cans, gloves, and hairnets may also find their way into by-product supplies. Quality control is usually not a concern for suppliers when providing waste products for cattle producers, so drastic variation can occur in resultant rations when utilizing by-product feeds.

Recommendations for Feeding Sweetpotato By-products

(1) Control variation in the ration. Use upper limits on feed ration ingredients and analyze each shipment of by-product feed from the supplier (Mehren 1996). Until further studies can be conducted, no more than 20% of the ration (dry matter basis) should contain sweetpotato or sweetpotato by-products. An exception would be a mix of 90% SPCW and 10% broiler litter fed free-choice as a slurry, which has proven to be an effective program (Rogers et al. 1999). Careful monitoring of moisture content and frequent ration adjustments may be necessary to ensure that cattle receive the desired nutrient intake (Crickenberger and Carawan 1991).

(2) Reduce incidence of choke/bloat. Chopping whole cull sweetpotatoes into smaller pieces (and fermenting if possible) and allowing to wilt or soften before feeding, will help in the prevention of choke and bloat. A rail or cable placed low over the feedbunk will also help keep cattles' heads down so that they are more prone to chew on the sweetpotato before swallowing (Mehren 1996).

(3) Avoid spoilage or freezing with high-moisture rations. High-moisture rations spoil very readily, especially in the heat of summer, and are prone to freeze solid in the bunk in colder climates. Feeding smaller amounts, but more frequently, should help control spoilage and freezing. The best time to begin feeding in areas of hot temperatures is very early morning and again at dusk, when temperatures are lowest.

(4) Adequately mix the ration. Often, rations contain several components that are difficult to handle and may not readily mix with other ingredients. Adequate mixing is necessary to provide ration consistency.

(5) Sweetpotatoes or their by-products should be used with caution in starting rations for stressed feeder cattle. These products generally may be utilized at or below 10% of the dry matter in a starting ration; however, caution should be exercised.

(6) Recommendations for storage and disposal of sweetpotatoes and their by-products.

- When deciding whether to feed sweetpotatoes or their by-products, keep several key concepts in mind. It is essential to determine the quantity of material available, the seasonality of the supply, the ability of the cattle producer to utilize available quantities, and whether feeding will benefit both the cattle producer and the processing plant (Crickenberger and Carawan 1991).
- Cull sweetpotatoes in storage should be clean. Culls should be washed before being ensiled, and ensiled sweetpotatoes should be incorporated into a balanced ration.
- Spread waste sweetpotatoes that cannot be fed to livestock on fields not intended for later production of this crop. Sweetpotatoes spread on the surface may regrow or contaminate adjacent sweetpotato crops. Spreading is best performed in the late fall or winter when tubers/roots will freeze or when insufficient rainfall will help to dry the sweetpotatoes. Sweetpotatoes should be disked into the soil to prevent runoff potential.
- Avoid large piles of sweetpotatoes and minimize any public nuisances. Large piles are likely to attract insects and produce objectionable odors; however, if large piles are unavoidable, locating them far from homes or businesses and spreading lime will help alleviate odor problems.

- Avoid drainage from sweetpotato storage areas. Runoff and drainage from sweetpotato storage sites should not contaminate any nearby waterways, wells, or groundwater (Glenn 1988).
- Fermented sweetpotato feeds should be stored in concrete-lined storage pits. Pits should be constructed in areas where there is no contamination or dilution from water runoff entering the pit. A structure (berm) should be constructed to prevent runoff of the waste in case an overflow of feed-mixing equipment should occur.

(7) Sample rations utilizing sweetpotatoes and their by-products. Sweetpotato by-products can generally be utilized in growing cattle rations at levels up to 20% of diet dry matter. Starter rations can include sweetpotatoes or sweetpotato by-products at levels up to 10% of the diet

Table 10.4 Fermented chopped sweetpotatoes (1/4-1")

Sample stocker complete growing ration utilizing 1/4-1" chopped cull sweetpotatoes (fermented two weeks), % on a dry matter basis (Poore et al. 1998) .

INGREDIENT	PERCENT
Cottonseed hulls	21
Bermudagrass haylage	21
Ground corn	32.1
Soybean meal	9.4
Sweetpotatoes	15
Limestone	1.0

Diet contains a premix included at 0.5% dry matter providing 30g lasalocid/ton dry matter, 1000 IU Vitamin A/lbs. dry matter, trace minerals, and salt.

Note: Above ration resulted in 2.07 lbs/day gain and 18.7 lbs dry matter intake/day on 500 to 700 lb heifers.

Table 10.5 Sweetpotato cannery solids

Sample ration utilizing mixture of SPCW and sweetpotato cannery solids, % given as both dry matter basis and as fed

INGREDIENT	PERCENT (DM)	PERCENT (AS FED)
SPCW plus sweetpotato cannery solids	19.92	45.04
Broiler litter	10.12	5.03
Winter pea/oat silage	11.92	15.00
Corn screenings/soyhulls	35.58	14.93
Cotton waste	12.00	5.03
Wet brewer's grains	8.16	14.00
Supplement (vitamin/mineral/ionophore)	2.30	0.96

Note: Maximum intake should be 40 to 45 lbs. for 500 lbs calves.
   Feed cost/lbs gain = $ 0.2809 at an ADG of 2 lbs/hd/day.

dry matter, and levels should be gradually increased up to 20% of the diet dry matter for growing cattle. Listed in tables 10.4 and 10.5 are sample rations utilizing some common by-products of the sweetpotato industries.

Summary

Animal nutritionists, veterinary practitioners, and other farm advisors have the opportunity to assist producers in lowering feed costs in certain regions of the country by incorporating sweetpotato by-products into safe and effective rations. A basic understanding of waste stream availability in specific regions, along with knowledge of the production and health aspects of these by-products is essential to ensure optimal economic and health performance. Moldy or rotten cull sweetpotatoes may cause acute interstitial pneumonia due to presence of the toxin 4-ipomeanol resulting in large herd losses. Caution should also be used when feeding sweetpotato cannery waste and other vegetable-processing slurries free-choice due to the potential for dental erosion.

Acknowledgments

The authors thank Norman Brown, Vice President of Bruce Foods Corporation, Wilson, NC; Bill Walter, Food Science Research Unit, Raleigh, NC; and the North Carolina Sweetpotato Commission for sharing their expertise on the sweetpotato processing industry and for their assistance with this manuscript.

References

ASHS. 1991. Publications Manual. Alexandria, VA, American Society for Horticultural Science.

Bath, D., J. Dunbar, J. King, S. Berry, R. Leonard and S. Olbrich. 1998. By-products and Unusual Feedstuffs. Feedstuffs Vol 70, No 30, pp. 32-38.

Briggs, G., W. D. Gallup, V. G. Heller, A. E. Darlow, and F. B. Cross. 1947. The Digestibility of Dried Sweet Potatoes by Steers and Lambs. Stillwater, OK, Oklahoma Agricultural Experiment Station, Oklahoma Agricultural and Mechanical College, Technical Bulletin No. T-28.

Crickenberger, R. and R. E. Carawan. 1991. Using Food Processing Byproducts for Animal Feed. Raleigh, NC, North Carolina Cooperative Extension Service Fact Sheet, #CD-37.

Dominguez, P. 1991. Feeding of sweet potato to monogastrics. Proc. Food Agricultural Organ. Expert Consultation, Food and Agricultural Organization, Cali, Columbia; pp. 217-233.

Doster, A., F. E. Mitchell, R. L. Farrell, and B. J. Wilson. 1978. Effects of 4-Ipomeanol, a product from mold-damaged sweetpotatoes, on the bovine lung. *Vet Pathol.* 15: 367-375.

Dukes, P. D., M. G. Hamilton, A. Jones, and J. M. Schalk. 1987. Sumor, a multiuse sweet potato. *HortScience* 22(1): 170-171.

Dunbar, J., A. Ahmadi, W. N. Garrett, J. W. Oltjen, and D. J. Drake. 1990. Taurus Least Cost Ration Analysis Programs for Beef Cattle. Dept. of Animal Science, University of California, Davis. Cooperative Extension computerized nutrition software *[http://animalscience.ucdavis.edu/extension/taurus.htm]*

Espinola, C. N. 1992. Alimentacion animal con batata (Ipomoea batatas) en Latinoamerica. Turrialba 42(1): 114-126.

FAO. 1986. Production Yearbook. Rome, Italy, Food and Agricultural Organization.

Glenn, R. 1988. Recycling waste potatoes on the farm. Misc. Rep. 318. Maine Agricultural Experiment Station, University of Maine, Augusta; pp. 1-5.

Greenough, P., F. J. MacCallum, and A. D. Weaver. 1981. Lameness in Cattle. Philadelphia, PA, J. B. Lipincott Co. pp. 219-227.

Hill, B. and H. F. Wright. 1992. Acute interstitial pneumonia in cattle associated with consumption of mold-damaged sweetpotatoes (Ipomoea batatas). *Aust Vet J* 60(2): 36-37.

Kummer, F. 1943. Equipment for Shredding Sweet Potatoes Prior to Drying for Livestock Feed. Auburn, AL, Agricultural Experiment Station of the Alabama Polytechnic Institute Circular. No. 89 pp. 1-13.

Mehren, M. 1996. Feeding Cull Vegetables and Fruit to Growing/Finish Cattle. Feeding Cull Vegetables and Fruit to Growing Cattle, Scottsdale, AZ, Academy of Veterinary Consultants Proceedings.

Morrison, F. 1946. Roots, tubers and miscellaneous forages. Feeds and Feeding. Ithaca, NY: The Morrison Publishing Company, page 312-315.

NCDA. 1996. Marketing North Carolina Sweetpotatoes Including Louisiana. North Carolina Department of Agriculture and Consumer Services, Raleigh, NC.

NCDA. 1998. North Carolina 1998 Agricultural Statistics. North Carolina Department of Agriculture and Consumer Services, Raleigh, NC. Publication No. 190.

Peckham, J., F. E. Mitchell, O. H. Jones, Jr., and B. Doupnik, Jr. 1972. Atypical interstitial pneumonia in cattle fed moldy sweetpotatoes. *JAVMA* 160(2): 169-172.

Peirce, L. 1987. Tuber and Tuberous Rooted Crops. Vegetables: Characteristics, Production, and Marketing. New York: John Wiley and Sons. pp. 287-308.

Poore, M. H., G. M. Gregory, J. L. Hart, and P. R. Ferket. 1998. Value of alternative ingredients in calf growing rations. *J. Anim. Sci.* 76 (Suppl. 1): 304 (abstr).

Radostits, O., D. C. Blood, and C. C. Gay. 1994. Veterinary Medicine A Textbook of the Diseases of Cattle, Sheep, Pigs, Goats and Horses. London, England, Bailliere Tindall, London, UK. pp. 1617-1621.

Rogers, P. . M. and D. B. R. Poole. 1987. Incisor wear in cattle on self-fed silage. *Vet Rec* 120:348.

Rogers, G. and M. H. Poore. 1994. Alternative feeds for reducing beef cow costs. *Vet Med* 89: 1073-1084.

Rogers, G., and M. H. Poore. 1997. Dental effects of feeding sweet potato cannery waste in beef cattle. *Compendium on Cont Ed for Practicing Vet* 19:541-546.

Rogers, G., M. H. Poore, B. L. Ferko, R. P. Kusy, T. G. Deaton, and J. W. Bawden. 1997. In vitro effects of an acidic by-product feed on bovine teeth. AJVR 58:498-503.

Rogers, G. M., M. H. Poore, B. L. Ferko, T. T. Brown, T. G. Deaton, and J. W. Bawden. 1999. Dental wear and growth performance in steers fed sweetpotato cannery waste. *JAVMA* 214:681.

Rolston, L. H., C. A. Clark, J. M. Cannon, W. M. Randle, E. G. Riley, P. W. Wilson, and M. L. Robbins. 1987. Beauregard sweetpotato. *HortScience* 22:1338-1339.

Ruiz, M. 1982. Sweet Potatoes (Ipomoea batatas [L] Lam) for Beef Production: Agronomic and Conservation Aspects and Animal Responses. Sweet Potato Proceedings of the First International Symposium, Shanhua, Tainan, Taiwan, Asian Vegetable Research and Development Center.

Sauter, E., D. D. Hinman, and J. F. Parkinson. 1985. The lactic acid and volatile fatty acid content and in vitro organic matter digestibility of silages made from potato processing residues and barley straw. *J Anim Sci* 60: 1087-1094.

Seath, D., L. L Rusoff, G. D. Miller, and C. Branton. 1947. Utilizing Sweet Potatoes as Feed for Dairy Cattle., Louisiana Agricultural Experiment Station, Bulletin #423.

Singletary, C., S. E. McCraine, and L. Berwick. 1950. Dehydrated Sweet Potato Meal for Fattening Steers. Louisiana State University and Agricultural and Mechanical College Bulletin #446, page 3-7.

Sistrunk, W., and I. K. Karim. 1977. Disposal of lye-peeling wastes from sweet potatoes by fermentation for livestock feed. *Arkansas Farm Res* 26(1): 8-9.

USDA. 1996. Nutrient Database for Standard Reference, Release 11, Sweetpotato, Raw.

Walter, Bill. Nutrient Content of Sweetpotatoes (July 10,1997), Telephone conversation.

Wilson, B., D. T. C. Yang and M. R. Boyd. 1970. Toxicity of mould-damaged sweet potatoes (Ipomoea batatas). *Nature* 227: 521-522.

Wilson, B. 1973. Toxicity of mold-damaged sweetpotatoes. *Nutrition Reviews* 31(3): 73-78.

Wilson, J. H., R. J. MacKay, H. Nguyen,and W. S. Cripe. 1981. Atypical Interstitial Pneumonia: Moldy Sweet Potato Poisoning in a Florida Beef Herd. *Florida Veterinary Journal* 10(Fall 1981): 16-17.

Wilson, L. and W. W. Collins. 1993. What is the difference between a sweetpotato and a yam? North Carolina Cooperative Extension Service.

Woolfe, J. 1992. Livestock feeding with roots and tubers. Sweetpotato: An Untapped Food Resource. Cambridge, England: Cambridge University Press.

Wu, J. 1980. Energy value of sweet potato chips for young swine. *J. Anim. Sci.* 51(6): 1261-1265.

11

The Use of Food Waste as a Feedstuff for Ruminants

by Paul Walker

Introduction

Diversity of agriculture in the United States and the proximity of the livestock and feeding industry to manufacturers of human foods provide every state a wide variety of by-product and nontypical feedstuffs for use in ruminant diets. In some instances, these feedstuffs can be economically used in beef, sheep, and dairy operations. However, many others are not worth the cost, labor, and added facilities required. The purpose of this chapter is to discuss the advantages and disadvantages of food waste as a feedstuff for ruminant animals.

Three reasons exist for feeding nontypical or by-product feeds. From the food processor's standpoint, ruminants can be an alternative method of waste disposal for unwanted residue as compared to discharging in a landfill area. Processors contract with vendors to dispose of unwanted residues, who then may contact livestock operations in an attempt to market these materials as ruminant feed. From the cattlemen's perspective, by-product feeds reduce ration costs and, thereby, increase profitability. Particularly for high-producing dairy cows, nontypical feedstuffs are utilized to increase nutrient density.

The meat-processing industry produces protein meals such as blood meal that are low in rumen degradability, and subsequently they can be used to supply the absorbable protein deficit in high-producing dairy cows. As a result of the Mammalian Protein Ruminant Feed Ban, many meat and bone meals have no or restricted uses in livestock diets. Nontypical or by-product feedstuffs such as discarded cereals, bakery products, and snack foods can be used to increase dietary energy content. The high lipid content of discarded snack foods increases calorie density without a subsequent reduction in milk fat content.

The primary reason for inclusion of nontypical feeds or food waste is to increase profitability by utilizing less expensive diets or increasing efficiency of existing dietary ingredients. The use of food waste by-products should be matched to animal requirements for specific production goals. Some food waste by-product feedstuffs are inappropriate in diets for dairy cattle but can be used in diets fed to beef cows. Unfortunately, what appears to be an inexpensive feedstuff, may in reality increase ration costs on a dry matter basis or per nutrient basis.

Several factors need to be considered before purchase of a nontypical food waste feedstuff. First, many of these by-product feedstuffs, especially fruit and vegetable wastes, contain high moisture levels and price evaluation must be performed on a dry matter basis. For instance, carrots offered at $10/ton delivered to a feedlot will contain approximately 88% water, which translates into a cost of $83.33/ton of dry matter. Even though carrots are nutritious feedstuffs, the amount of moisture present may limit their usefulness. Feeding high- moisture by-products increases both the amount of feed delivered to the bunk and the management required to ensure fresh feed is available at all times. Consequently, feedbunks may have to be larger and cleaned more frequently. This will be especially important during the summer months. Generally, vegetarian food waste (food waste devoid of meat or animal protein) is delivered to the farm in truckload lots, which requires adequate storage and handling facilities. High moisture feedstuffs can deteriorate rapidly during warm weather, which will reduce palatability and quality. High moisture food waste may require special handling facilities and containment areas to minimize runoff.

Food waste or nontypical feedstuffs occasionally have excessive levels of certain minerals or may be contaminated with pesticides. For instance, corn gluten feed may contain high levels of sulfur that may induce polioencephalomalacia (PEM or "Polio") when fed in high proportions of the diet. The high sulfur content that results from chemicals added during the extraction procedure may decrease the availability of thiamine, which results in PEM. Likewise, potato screenings and peels may contain excessive levels of sodium that will limit the effectiveness of monensin (Ely Lily Co., Indianapolis, IN), a commonly used feed additive that improves feed to gain ratios in beef cattle diets, and reduce dry matter consumption when fed at high levels. In addition, certain food waste has low levels of calcium or high levels of phosphorus that may cause an imbalance in the calcium to phosphorus ratio. The minimum recommended relationship of calcium to phosphorus should be in excess of 1.5 to 1. Dietary phosphorus levels in excess of calcium can cause urinary calculi. Additional calcium needs to be added to diets containing high levels of corn gluten feed because of the high phosphorus level. Some food waste contains high levels of salt. Excess sodium in the diets of cattle can result in reduced feed intakes, lower performance, and in extreme cases nervous disorders, hypertension, and blindness. Generally the sodium levels found in food waste can be accommodated in balanced diets without detriment to cattle.

Fruit- and vegetable-processing waste may contain pesticide residues that can have adverse effects on cattle. Processing wastes that are suspect for high pesticide levels should be destroyed or ensiled and tested before feeding. Another safety concern is the potential for salmonellosis and listeriosis. Salmonellosis may be contracted from fresh, unprocessed vegetable wastes and green chop forages. Documented cases of salmonellosis have been reported as the result of feeding fresh green chop forages. Listeriosis occurs from feeding decomposing plant materials. Both of these diseases are of concern because of the potential health hazards from contaminated meat or milk. These safety concerns should not prevent the use of food waste by-products in ruminant diets. However, appropriate management procedures should be initiated to minimize the potential hazards.

Another concern with nontypical feedstuffs is the daily variation in nutritive content. This variation occurs between batches from the same manufacturing plant and, also, between plants. Variation in moisture content can alter the economic value of the feedstuff and palatability. The vendor or manufacturer should provide a minimum guarantee on moisture and nutritive content or should sufficiently lower the price to accommodate the variation. Periodic documentation of the moisture and nutritive contents may be necessary to ensure the guarantee. These details can be negotiated for each vendor-client relationship.

The decision to purchase any food waste by-product and nontypical feedstuffs should be based on the relative value of feedstuffs currently utilized. For most operations, these feedstuffs would be priced against alfalfa, corn, and soybean meal. The nutritive contents of many by-product and nontypical feedstuffs are shown in table 11.1. Values in this table are average values and may not represent all feedstuffs accurately. If there is any doubt, nutrient analysis should be performed. Nutrient and mineral content can vary as well. In a recent survey conducted in Missouri of several processing plants, coefficients of variation for acid detergent fiber (ADF), neutral detergent fiber (NDF), ash, in vitro dry

Table 11.1. Nutrient composition of various by-product and nontypical feedstuffs (dry matter basis)

Feedstuffs	Dry Matter (%)	Metabolizable Energy (Mcal/lb)	Net energy Maintenance (Mcal/lb)	Net energy gain (Mcal/lb)	Crude protein (%)	Ether extract (%)	Ash (%)	Calcium (%)	Phosphorus (%)	Potassium (%)
Animal fat	99	2.91	2.16	1.60	0.0	99.5	0.0	0.00	0.00	0.00
Apple pomace	40	0.92	0.54	0.28	5.6	5.2	3.5	0.13	0.12	0.49
Bakery waste	95	1.55	1.06	0.73	11.2	12.7	4.4	0.14	0.26	0.53
Beet pulp	91	1.22	0.80	0.52	9.7	0.6	5.4	0.69	0.10	0.20
Beet tailings	15	0.84	0.45	0.20	6.8	2.8	32.5	1.56	0.29	5.74
Blood meal	92	1.08	0.64	0.37	86.0	1.4	5.8	0.32	0.26	0.10
Brewers grains	23	1.09	0.69	0.41	27.0	6.5	4.8	0.33	0.55	0.09
Brewers grains	91	1.09	0.69	0.41	27.2	7.2	3.9	0.33	0.55	0.09
Broiler litter	89	1.09	0.69	0.41	24.5	3.0	22.0	3.16	1.78	1.68
Buckwheat	88	1.18	0.77	0.49	12.5	2.8	2.3	0.11	0.37	0.51
Canola meal	90	1.25	0.83	0.54	36.5	7.9	7.5	0.72	1.14	0.90
Carrots	12	1.38	0.94	0.64	9.9	1.4	8.2	0.40	0.35	2.80
Cookie meal	90	1.52	1.05	0.73	7.0	14.0	4.4	0.14	0.26	0.53
Corn	87	1.48	1.02	0.70	9.5	4.2	1.4	0.02	0.35	0.37
Corn cobs	90	0.82	0.44	0.19	3.2	0.7	1.7	0.12	0.04	0.87
Corn gluten feed	40	1.36	0.92	0.62	26.2	2.4	7.5	0.36	0.82	0.64
Corn gluten feed	92	1.36	0.92	0.62	26.2	2.4	7.5	0.36	0.82	0.64
Corn gluten meal	90	1.46	1.00	0.69	67.2	2.4	1.8	0.08	0.54	0.21
Corn silage	35	1.15	0.74	0.47	8.0	3.1	4.5	0.23	0.22	0.96
Cottonseed hulls	91	0.69	0.31	0.07	4.1	1.7	2.8	0.15	0.09	0.87
Cottonseed meal	90	1.31	0.83	0.54	44.0	1.6	7.1	0.18	1.21	1.52
Cottonseeds	93	1.58	1.10	0.77	23.9	23.1	4.8	0.16	0.75	1.21
Cull beans	90	1.38	0.94	0.64	25.3	1.5	5.2	0.18	0.59	1.47
Cull cucumbers	8	0.95	0.64	0.27	17.9	8.9	8.9	0.24	0.43	3.32
Distillers grain	93	1.41	0.96	0.66	23.0	9.8	2.4	0.11	0.43	0.18
Feather meal	93	1.15	0.74	0.47	91.3	3.2	3.8	0.28	0.72	0.31
Fish meal	90	1.22	0.76	0.48	67.0	8.0	21.0	5.90	3.30	0.60

Table 11.1 (*continued*)

Grape pomace	20	0.45	0.05	0.00	13.0	7.9	10.3	0.34	0.12	0.35
Hominy feed	90	1.55	1.02	0.75	11.5	7.7	3.1	0.05	0.57	0.65
Lettuce	5	0.84	0.38	0.08	16.6	4.1	15.9	0.86	0.46	4.52
Linseed meal	90	1.28	0.85	0.56	39.8	1.5	6.5	0.43	0.89	1.53
Malt sprouts	94	1.17	0.70	0.43	28.1	1.4	7.0	0.23	0.75	0.23
Meat scraps	94	1.17	0.70	0.43	54.8	9.7	23.4	6.37	3.33	0.60
Meat/bone meal	93	1.09	0.65	0.40	50.4	10.4	31.5	11.06	5.48	1.43
Mint by-products	27	0.94	0.55	0.21	14.0	1.8	16.0	1.10	0.57	0.00
Molasses	78	1.30	0.87	0.58	8.5	0.2	11.3	0.17	0.03	6.07
Oat hulls	92	0.58	0.19	0.00	3.9	1.8	6.6	0.15	0.15	0.62
Oat screenings	90	1.26	0.78	0.54	12.9	4.6	2.5	0.08	0.49	0.55
Peas	89	1.43	0.98	0.67	25.3	1.4	3.3	0.15	0.44	1.13
Potato by-products	53	1.43	0.99	0.68	5.3	0.4	3.4	0.04	0.18	1.38
Potato screenings	33	1.33	0.92	0.64	8.4	0.4	3.4	0.16	0.25	0.39
Rice hulls	92	0.20	0.00	0.00	3.3	0.8	20.6	0.10	0.08	0.57
Rye	90	1.38	0.94	0.64	13.8	1.7	1.9	0.07	0.37	0.52
Safflower meal	92	0.94	0.79	0.50	25.4	1.4	8.2	0.37	0.81	0.82
Soy hulls	90	1.08	0.65	0.39	8.0	2.1	5.1	0.49	0.21	1.27
Soybean meal	89	1.38	0.94	0.64	44.0	1.5	7.3	0.33	0.71	2.14
Speltz	90	1.23	0.81	0.53	13.3	2.1	3.9	0.13	0.42	0.50
Sunflower meal	93	1.07	0.67	0.40	49.8	3.1	8.1	0.44	0.98	1.14
Sweet corn waste	32	1.18	0.77	0.49	7.7	5.2	4.9	0.30	0.90	1.15
Thin stillage	5	1.45	0.99	0.68	29.7	9.2	7.8	0.35	1.37	1.80
Triticale	90	1.38	0.94	0.64	17.6	1.7	2.0	1.70	0.27	1.92
Vegetable fat	100	2.91	2.16	1.60	0.0	99.9	0.0	0.00	0.00	0.00
Wheat by-products	90	1.14	0.73	0.45	18.8	4.9	5.2	0.13	0.99	1.13
Whey	7	1.55	1.09	0.75	13.0	4.3	8.7	0.73	0.65	2.75
Whole soybeans	92	1.50	1.03	0.71	42.8	18.8	5.5	0.27	0.65	1.82

Source: Presented by Steven Rust (Michigan State University) at the 1991 Minnesota Nutrition Conference.

matter digestibility (IVDMD), crude protein, fiber bound nitrogen (ADIN), starch, and fat analyses of five by-product feeds (corn gluten feed, dry distillers grain, rice hulls, soybean hulls, and whole soybeans) were 6.0, 4.7, 40.0, 1.7, 4.6, 30.5, 16.1, and 18.1% respectively. The components with the greatest variation were energy (fat and starch), mineral content (ash), and nitrogen availability (ADIN).

Beef Cattle and Sheep

The values of various food waste by-product or nontypical feeds in different types of ruminant diets are presented in tables 11.2 and 11.3. Table 11.2 presents calculated values (as-fed) for the various by-products based on corn silage priced at $18.90/ton and soybean meal priced at $200 or $265/ton. The comparisons in table 11.2 are based on the ratio of metabolizable energy values for the by-product feed and corn silage in growing diets for beef cattle and sheep. Monetary adjustments are made to correct for additional protein needed or saved. Notice the column heading "maximum recommended feeding level." In very few instances can by-product feeds replace all of the corn silage in the diet without severe reductions in performance. The column with the 8% crude protein heading, provides a price comparison for the various feedstuffs based on the proportion of metabolizable energy content present in corn silage. Adjustments for crude protein content are not included in this column. Since many of the by-products contain crude protein values that are different from that of corn silage and because of the added expense of supplemental protein (generally soybean meal), adjustments for crude protein deficit or excess were performed. The value of the nontypical feedstuffs will also be influenced by the different supplemental protein costs and crude protein levels in the diet. Cost per unit of crude protein of $3/% unit, $4.5/% unit, and $6/% unit are equivalent to the following soybean meal and urea prices: 50% of the crude protein from urea ($180/ton) and 50% from soybean meal ($200/ton); 100% soybean meal ($200/ton); and 100% soybean meal ($265/ton), respectively. The cost per percentage unit of crude protein is calculated by dividing the price of the protein source by its protein content. For example, soybean meal (44% CP) priced at $200/ton would have a cost per percentage unit of crude protein or $4.54 ($200/44%). Two dietary protein levels (11 and 13%) are presented under each cost per unit of crude protein heading. To fully understand the table, consider apple pomace. The maximum recommended feeding level is 25% of the ration dry matter. It has 80% of the metabolizable energy (ME) value of corn silage and, on an ME basis, is worth $17.33/ton (as-fed) when corn silage is worth $18.90/ton. Since apple pomace has less crude protein than corn silage, additional supplemental protein is required, which reduces its value. If the supplemental protein is a urea-soybean meal mixture with a cost of $3 per percentage unit of crude protein, apple pomace would be worth $14.45/ton delivered to the feedbunk. A supplemental crude protein cost of $6.00 per percentage unit would reduce the value to $11.57/ton (as-fed). Corn gluten feed has more crude protein than corn silage and, thus increases its value over the metabolizable energy supplied. In this case, on an ME basis only, corn gluten feed (40% DM) is worth $25.61. However, credit for the additional crude protein increases the value to $29.21/ton in an 11% crude protein ration or $31.61 in a 13% crude protein ration when compared at a supplemental crude protein cost of $3/% unit. The value of vegetable-processing wastes may not be sufficient to return transportation costs in many instances.

To compare the value of a by-product or nontypical feedstuff in a maintenance type diet (beef cows), utilize the value listed in the 8% crude protein column in table 11.2.

Table 11.3 compares the value of food waste by-product and nontypical feedstuffs that can replace corn in finishing diets for beef cattle and sheep. The table format is identical to table 11.2 and can be utilized in a similar manner. Corn was priced at $2.00/bushel. Animal fat has 197% (NRC 1984) the ME value of corn and should not exceed 5% of the ration dry matter. On an energy basis only, animal fat would be worth $161.81/ton (or $0.08/lb) as a replacement for corn priced at $2.00/bushel (NRC 1984, ME value for fat may be an underestimate). However, animal fat is devoid of crude protein, and supplemental protein would have to be added to equalize the crude protein value of corn. In this case, with $3.00 per percentage unit of supplemental crude protein, animal fat is worth $28.21/ton less ($133.60/ton). The appropriate value to utilize for animal fat could depend on the other factors such as the amount of heat stress and dustiness of the diet. Diets that exceed (greater than 12.0%) crude protein requirements would benefit from 5% animal fat in the diet without causing a protein deficit. The price relationships presented in tables 11.2 and 11.3 would change as least cost ration formulation techniques are utilized.

Table 11.2. Value ($/ton) of various by-product and nontypical feedstuffs in growing rations for beef cattle and sheep

Feedstuffs	Dry matter (%)	Metabolizable energy (Mcal/lb)	Energy conversion factor	Crude protein (%)	Maximum recommended dietary level (% of DM)	8[b,c] (%)	Cost of supplemental crude protein ($/% unit)[a]					
							3		4.5		6	
							11[b,d]	13[b,d]	11[b,d]	13[b,d]	11[b,d]	13[b,d]
							Value ($/ton [as-fed])					
Apple pomace	40	0.92	0.80	5.6	25	17.33	14.45	14.45	13.01	13.01	11.57	11.57
Beet pulp	91	1.22	1.06	9.7	25	52.05	56.69	56.69	59.01	59.01	61.34	61.34
Beet tailings	15	1.15	1.00	6.8	20	8.07	7.53	7.53	7.26	7.26	6.99	6.99
Broiler litter	40	1.09	0.94	24.5	35	20.40	24.00	26.40	25.80	29.40	27.60	32.40
Buckwheat	88	1.18	1.03	12.5	35	48.83	56.75	60.71	60.71	66.65	64.67	72.59
Carrots	12	1.38	1.20	9.9	20	7.79	8.47	8.47	8.81	8.81	9.15	9.15
Corn cobs	90	0.82	0.72	3.2	15	34.77	21.81	21.81	15.33	15.33	8.85	8.85
Corn gluten feed	40	1.36	1.19	26.2	90	25.61	29.21	31.61	31.01	34.61	32.81	37.61
Corn gluten feed	92	1.36	1.19	26.2	50	58.91	67.19	72.71	71.33	79.61	75.47	86.51
Corn silage	35	1.15	1.00	8.0	90	18.90	18.90	18.90	18.90	18.90	18.90	18.90
Cottonseed hulls	91	0.69	0.60	4.1	40	29.52	18.88	18.88	13.55	13.55	8.23	8.23
Cull beans	90	1.38	1.20	25.3	25	58.40	66.50	71.90	70.55	78.65	74.60	85.40
Cull cucumbers	8	0.95	0.83	17.9	25	3.59	4.31	4.79	4.67	5.39	5.03	5.99
Grape pomace	20	0.45	0.39	13.0	20	4.18	5.98	7.18	6.88	8.58	7.78	10.18
Lettuce	5	0.84	0.73	16.6	30	1.97	2.42	2.72	2.65	3.10	2.87	3.47
Mint by-products	27	0.94	0.82	14.0	25	11.93	14.36	15.98	15.57	18.00	16.79	20.03
Oat hulls	92	0.58	0.50	3.9	20	24.94	13.62	13.62	7.96	7.96	2.31	2.31
Oat screenings	90	1.26	1.10	12.9	40	53.40	61.50	66.63	65.55	73.25	69.60	79.86
Rice hulls	92	0.20	0.17	3.3	15	8.44	-4.53	-4.53	-11.01	-11.01	-17.50	-17.50

Table 11.2. (*continued*)

Soy hulls	90	1.05	0.91	8.0	80	44.37	52.06	17.76	3.42	48.02	3.51	64.60
Speltz	90	-1.23	1.07	13.3	50	44.37	60.16	17.47	3.87	56.12	4.14	72.88
Sweet corn waste	32	1.18	1.03	7.7	80	44.37	65.56	17.47	4.17	61.52	4.56	78.40
Thin stillage	5	1.45	1.26	29.7	10	44.37	64.21	17.33	4.09	60.17	4.46	77.02
Wheat by-products	90	1.14	0.99	18.8	35	44.37	72.31	17.33	4.54	68.27	5.09	85.30
Whey	7	1.07	0.93	13.0	25	44.37	68.26	17.18	4.32	64.22	4.77	81.16
Whole soybeans	92	1.50	1.30	42.8	25	44.37	79.06	17.18	4.92	75.02	5.61	92.20

Source: Presented by Steven Rust (Michigan State University) at the 1991 Minnesota Nutrition Conference.

[a] Cost/% unit of protein is calculated by division of ingredient cost/ton by crude protein content. $3.00/% unit = 50% crude protein equivalent from urea ($180/ton) and 50% from SBM ($200/ton); $4.50/% unit = $200/ton for SBM; $6.00/% unit = $265/ton for SBM. Corn silage was priced at $18.90/ton.

[b] Dietary crude protein level.

[c] Value of each feedstuff based on energy content only.

[d] Value of each feedstuff based on energy content of corn silage and added SBM to match crude protein in corn silage or credit feedstuff for crude protein content in excess of corn silage.

Calculations

A. Corn silage price − $18.90/ton (35% DM).

B. Corn silage price on a dry matter basis − $54.00/ton ($18.90/.35).

C. Price of alternative feedstuff (feed x) with crude protein content less than corn silage.
[($54.00 × energy conversion factor) + (crude protein content of feed x − 8) × (cost per unit of supplemental crude protein)] ÷ (dry matter content of feed x/100).

D. Price of alternative feedstuff (feed y) with crude protein content greater than desired level in ration.
[($54.00 × energy conversion factor) + (desired crude protein in ration − 8) × (cost per unit of supplemental protein)] ÷ (dry matter content of feed y/100).

Table 11.3. Value ($/ton) of various by-product and nontypical feedstuffs in growing rations for beef cattle and sheep

Feedstuffs	Dry matter (%)	Metabolizable energy (Mcal/lb)	Energy conversion factor	Crude protein (%)	Maximum recommended dietary level (% of DM)	Value ($/ton [as-fed])						
						9.5[b,c] (%)	Cost of supplemental crude protein ($/% unit)[a]					
							3		4.5		6	
							11[b,d]	12[b,d]	11[b,d]	12[b,d]	11[b,d]	12[b,d]
Animal fat	99	2.91	1.97	0.0	5	161.81	133.60	133.60	119.49	119.49	105.38	105.38
Bakery waste	95	1.55	1.05	11.2	20	82.49	86.76	87.33	88.90	89.76	91.04	92.18
Beet pulp	91	1.22	0.82	9.7	25	62.28	62.83	62.83	63.10	63.10	63.38	63.38
Brewers grains	23	1.09	0.74	27.0	30	14.04	15.07	15.76	15.59	16.63	16.11	17.49
Brewers grains	91	1.09	0.74	27.2	30	55.54	59.64	62.37	61.69	65.78	63.73	69.19
Buckwheat	88	1.18	0.80	12.5	35	58.43	62.39	65.03	64.37	68.33	66.35	71.63
Carrots	12	1.38	0.94	9.9	20	9.32	9.46	9.46	9.53	9.53	9.60	9.60
Cookie meal	90	1.52	1.03	7.0	25	77.00	70.25	70.25	66.87	66.87	63.50	63.50
Corn	87	1.48	1.00	9.5	100	72.21	72.21	72.21	72.21	72.21	72.21	72.21
Corn cobs	90	0.82	0.56	3.2	15	41.60	24.59	24.59	16.09	16.09	7.58	7.58
Corn gluten feed	40	1.36	0.92	26.2	90	30.65	32.45	33.65	33.35	35.15	34.25	36.65
Corn gluten feed	92	1.36	0.92	26.2	50	70.49	74.63	76.70	76.70	80.84	78.77	84.29
Cull beans	90	1.38	0.94	25.3	25	69.87	73.92	76.62	75.95	80.00	77.97	83.37
Distillers grain	93	1.41	0.96	23.0	60	73.86	78.05	80.84	80.14	84.33	82.23	87.81
Hominy feed	90	1.55	1.05	11.5	30	78.15	82.20	83.55	84.22	86.25	86.25	88.95
Malt sprouts	94	1.17	0.79	28.1	30	61.94	66.17	68.99	68.28	72.51	70.40	76.04
Molasses	78	1.30	0.88	8.5	20	56.97	54.63	54.63	53.46	53.46	52.29	52.29
Oat screenings	90	1.26	0.86	12.9	40	63.90	67.95	70.65	69.97	74.00	72.00	77.40
Peas	89	1.43	0.97	25.3	25	71.60	75.60	78.27	77.60	81.61	79.61	84.95
Potato by-products	53	1.43	0.97	5.3	15	42.64	35.96	35.96	32.62	32.62	29.28	29.28
Potato by-products	33	1.33	0.90	8.4	15	24.69	23.60	23.60	23.06	23.06	22.52	22.52

194

Table 11.3. *(continued)*

Rye	90	1.38	13.8	50	69.87	73.92	76.62	75.95	80.00	77.97	83.37
Speltz	90	1.23	13.3	50	62.29	66.34	69.04	68.36	72.41	70.39	75.79
Thin stillage	5	1.45	29.7	10	4.09	4.31	4.46	4.42	4.65	4.54	4.84
Triticale	90	1.38	17.6	90	69.87	73.92	76.62	75.95	80.00	77.97	83.37
Vegetable fat	100	2.91	00.0	5	163.45	134.92	134.95	120.70	120.70	106.45	106.45
Wheat by-products	90	1.14	18.8	35	57.46	61.51	64.21	63.54	67.59	65.66	70.96
Whey	7	1.07	13.0	25	4.20	4.52	4.73	4.67	4.99	4.83	5.25
Whole soybeans	92	1.50	42.8	25	77.30	81.44	84.20	83.51	87.65	85.58	91.10

Source: by Steven Rust (Michigan State University) at the 1991 Minnesota Nutrition Conference.

[a] Cost/% unit of protein is calculated by division of ingredient cost/ton by crude protein content. $3.00/% unit = 50% crude protein equivalent from urea ($180/ton) and 50% from SBM ($200/ton); $4.50/% unit = $200/ton for SBM; $6.00/% unit = $265/ton for SBM. Corn silage was priced at $18.90/ton.

[b] Dietary crude protein level.

[c] Value of each feedstuff based on energy content only.

[d] Value of each feedstuff based on energy content of corn silage and added SBM to match crude protein in corn silage or credit feedstuff for crude protein content in excess of corn.

Calculations

A. Corn price − $2.00/bushel (87% DM).

B. Corn price − dry matter basis − $82.00/ton($2.00/bushel × $\dfrac{36.7\ \text{bushel}}{\text{ton}}$) ÷ 0.87.

C. Price of alternative feedstuff (feed x) with crude protein content less than corn silage.
[($54.00 × energy conversion factor) + (crude protein content of feed x − 9.5) × (cost per unit of supplemental crude protein)] ÷ (dry matter content of feed x/100).

D. Price of alternative feedstuff (feed y) with crude protein content greater than desired level in ration.
[($54.00 × energy conversion factor) + (desired crude protein in ration − 9.5) × (cost per unit of supplemental protein)] ÷ (dry matter content of feed y/100).

195

Dairy Cattle

Food waste by-product or nontypical feeds are added to lactating dairy cattle diets to enhance caloric density, increase absorbable protein flow to the small intestine, reduce rumen acidosis, and finally, to lower feed costs. High-producing dairy cows are unable to consume sufficient energy during early lactation to meet requirements. Addition of fats or nontypical feeds high in fat, such as discarded snack foods and processed food waste, can be used to increase caloric density in diets in early lactation. Similarly, in early lactation, absorbable protein delivered to the small intestine or escape protein may be limiting production. Appropriately processed animal protein meals and by-products from the swine packing industry can be added to increase escape protein, and subsequently increase milk production. Feedstuffs that are high in solubles and low in fiber, such as corn silage or lush haylage, can result in low rumen pH and acidosis. Addition of food waste by-product or nontypical feedstuffs to slow the rate of digestion or stimulate salivary production and gut motility in these types of diets has been useful. Lastly, by-product or nontypical feedstuffs are added to diets to lower feed costs. The calculation of value should be based on the protein and energy that can be replaced by the feedstuff in question. This is best determined by least cost ration formulation. However, general rules of thumb can be generated based on crude protein and energy replaced. The nutritive values and recommended inclusion rates for various by-product or nontypical feedstuffs are shown in table 11.4. A value for a specific food waste or nontypical feedstuff can be calculated by the use of energy and protein factors (table 11.5). To determine the value of a feedstuff, multiply the energy factor from table 11.5 by the price of corn (\$/cwt) and add to it the product of protein factor times the price of soybean meal (\$/cwt). For example, the value of corn gluten meal on a dry matter basis is \$21.39/cwt [(\$4.40/cwt) × (−0.272)] + [(\$16.00/cwt) × 1.412] when corn is priced at \$4.40/cwt and soybean meal at \$16.00/cwt.

Comparison of protein sources is shown in table 11.6. Price comparisons are based on the energy and protein value of soybean meal. Protein sources with less metabolizable energy content than soybean meal were discounted using the energy value and price of corn (\$2.00/bushel) to equalize energy content. Again, notice the recommended maximum level in the diet. Blood meal is worth \$156.67 and \$480.34/ton when soybean meal costs are \$80 and \$240/ton, respectively.

Food By-products and Food Waste As Animal Feed

Much work has been done in researching the feeding of food by-products and food waste to cattle and sheep. The particular food by-products and waste fed to livestock vary with the region of the country and are dependent on the food products that are manufactured in the area or on the crops grown there. In most areas of the country, using food by-products and/or vegetables that do not meet market standards as food for ruminants is a practice that has been used with some success for many years. Long before any research was done on the value of by-products and unusual feedstuffs as part of the livestock ration of ruminants, farmers were feeding their cattle and sheep vegetable waste such as over-ripe apples or substandard vegetables. Carrots and cantaloupes that do not make the vegetable packers' grade have been used as a successful part of the diet of cattle in the southwestern United States. The high moisture content of the carrots and cantaloupes helps keep cattle from dehydrating in the dry climate, and cattle accept the new feed well because they enjoy the sweetness of the vegetables. However, because of seasonality, the carrots and cantaloupe are available only in the late spring and early summer. In addition, many farmers would rather disk them into the soil for use as a fertilizer than try to sell and transport them because the prices feedlot managers are able to pay for the substandard fruit and vegetables are so low. Feedlot managers in the Midwest have experimented with everything from apple- to potato-processing residue to french fries and salad dressing. The Midwest's proximity to manufacturers of human food products like cereals, breads, and snack foods made it one of the first areas in the country to experiment with the use of by-products as animal feed. By-products of the candy, cotton, vegetable, and citrus industries have also been used as livestock feed in the South and West. Hawaiian cattle ranchers have used a pineapple bran consisting of the outer shell of the pineapple and pineapple cannery waste products as a feed component.

Fresh Pulped Food Waste

In the fall of 1993, Illinois State University, Normal (ISU), began recycling cafeteria food waste as a feedstuff for ruminants. ISU remodeled four of its residence hall cafeterias to accommodate one or more Hobart (Hobart Co., Troy, OH) waste pulpers. These cafeterias serve upward to 8,000 students one or more times daily. Waste pulpers are wet grinders originally developed to reduce the volume, as opposed to weight, of waste paper going to landfills when tipping fees were based on cubic

Table 11.4. Composition and relative costs of protein and energy from selected by-product and nontypical feeds for dairy cattle

	Dry Matter (%)	NE1 3 × M (Mcal/lb DM)	Crude protein (adjusted) (% DM)	Acid detergent fiber (% DM)	Ca (% DM)	P (% DM)	Mg (% DM)	K (% DM)	S (% DM)	Maximum % of total ration DM
Alfalfa	88	.58	16	38	1.18	.27	.24	1.70	.19	80
Apple pomace	32	.40	4	41	.17	.11	.07	.40	—	25
Bakery waste	91	.97	9	2	.07	.11	—	—	—	20
Barley	88	.87[a]	14	7	.05	.37	.15	.45	.17	60
Beans, field, navy (cull)	88	.85	23	6	.17	.58	.15	1.50	.26	15
Beet pulp (sugar)	20	.75	8	29	4.20	.29	.24	.19	.20	20
Beet tailings (sugar)	15	.60	12	25	—	—	—	—	—	20
Brewers grains, wet	22	.78[b]	25[c]	25	.25	.54	.16	.09	.25	30
Brewers grains, dry	92	.75[b]	24[c]	29	.25	.54	.16	.09	.25	25
Brewers yeast	15	.82	48	4	.29	2.06	.31	2.40	.40	10
Candy-salvage	95	.90	0	0	—	—	—	—	—	5
Canola meal	92	.72	39	11	.75	1.28	.25	1.40	—	—
Citrus pulp	90	.88	6	22	2.00	.13	.16	.77	.07	25
Corn gluten feed	90	.81[b]	23	12	.10	1.00	.51	1.50	.40	25
Corn gluten meal	91	.93[b]	66	4	.08	.51	.09	.20	.42	20
Corn	88	.93[a]	10	3	.02	.26	.14	.50	.08	60
Corn silage	35	.70	8	28	.30	.20	.17	1.00	.07	45
Cottonseeds	91	.96[b]	25	32	.15	.73	.35	.73	.26	15
Cottonseed meal	92	.83	45	20	.17	1.07	.59	1.53	.28	25
Distillers grains	90	.87[b]	24[c]	16	.30	1.40	.50	1.80	.40	25

198

Table 11.4. (continued)

										2
Fat	100	2.35	—	—	—	—	—	—	—	—
Grape pomace	90	.26	—	54	—	—	—	—	—	—
Grain dust	88	.90	6	4	.10	.18	.26	.40	.03	30
Hominy	91	.95[b]	12[c]	13	.01	.58	1.20	.60	.30	10
Meat and bone meal	90	.74	54	—	10.30	5.40	.40	1.40	.40	10
Molasses, cane	75	.78	4	0	1.00	.10	.10	3.00	.23	20
Oats	88	.79	14	17	.07	.38	.14	.40	—	20
Potato peelings	30	.78	8	7	.10	.19	.10	.31	.07	20
Potato screenings	40	.82	7	3	.24	.19	.09	.67	.09	15
Potato by-product	55	.94	8[c]	—	.04	.22	.12	2.00	—	—
Screenings (corn-cereal)	88	.84	13	11	.30	.30	.10	—	.47	35
Soybean meal	90	.82	49	10	.34	.70	.30	2.20	.24	25
Soybean	89	1.00	42	7	.27	.65	.29	1.80	—	10
Soy hulls	90	.78[b]	12[c]	46	.45	.17	—	1.00	—	20
Steep water, corn	50	.90	0	0	—	—	—	—	.15	10
Wheat	89	.94[a]	14	4	.05	.40	.13	.40	.18	40
Wheat mids or mill run	90	.85	18	10	.11	.96	.39	1.10	.37	20
Whey	7	.81	13	0	.90	.81	.14	2.30	—	20

Source: Dairy Nutrition Handbook, Michigan State University (1990).

[a] Cracked = .84.

[b] NE$_l$ value may be less than stated when fed to high producing cows and especially when larger quantities are fed.

[c] New crude protein corrected for unavailable protein.

Table 11.5. Factors for comparison of by-product and nontypical feedstuffs for lactating dairy cows based on replacement value of shelled corn and 44% soybean meal (100% DM basis)

Grains	Net energy for lactation (Mcal/lb)	Crude Protein (%)	Corn factor	Soy factor	Forage and substitutes	Net energy for lactation (Mcal/lb)	Crude protein (%)	Corn factor	Soy factor
Bakery waste	0.93	11.9	0.981	0.048	Alfalfa hay, early	0.61	20.0	0.130	0.253
Barley	0.87	13.9	0.811	0.098	Alfalfa hay, mid	0.59	18.0	0.130	0.202
Beet pulp	0.81	8.0	0.809	-0.051	Alfalfa hay, late	0.51	15.5	0.013	0.139
Brewers grain, dry	0.68	27.1	0.127	0.432	Alfalfa-grass hay	0.53	14.0	0.087	0.101
Brewers grain, wet	0.72	28.0	0.193	0.455	Apple pomace	0.71	4.9	0.662	-0.129
Canola meal	0.76	42.0	-0.019	0.808	Beet mangels	0.81	11.4	0.737	0.035
Canola, whole	1.21	21.5	1.372	0.290	Corn cobs, ground	0.47	2.8	0.197	0.182
Citrus pulp	0.80	6.9	0.811	-0.078	Corn stalks	0.43	5.9	0.047	-0.104
Corn	0.92	10.0	1.000	0.000	Corn silage, eared	0.72	8.0	0.618	0.051
Corn gluten meal	0.88	65.9	-0.272	1.412	Corn silage, no ears	0.60	5.9	0.408	-0.104
Corn gluten feed	0.86	25.0	0.554	0.379	Corn silage, NPN	0.69	12.5	0.459	0.063
Corn, ground ear	0.84	9.3	0.845	-0.018	Corn, sweet cannery waste	0.72	8.8	0.601	-0.030
Cottonseed meal	0.77	44.8	-0.057	0.879	Grass hay	0.56	11.0	0.214	0.025
Cottonseed, whole	1.03	24.9	0.917	0.376	Oat straw	0.48	4.4	0.185	-0.141
Cull beans	0.90	22.9	0.684	0.326	Potato waste	0.82	9.6	0.796	-0.010
Distillers grain	0.88	29.5	0.501	0.492	Small grain silage	0.56	10.0	0.236	0.000
Fat	2.28	0.0	4.100	-0.253	Sorghum x sudan	0.60	11.0	0.299	0.025
Hominy	0.97	11.8	1.068	0.045	Soy straw	0.44	5.2	0.083	-0.121
Linseed meal	0.81	38.6	0.159	0.722	Wheat straw	0.45	4.2	0.125	-0.146
Oats	0.79	13.6	0.648	0.091	Wheat straw, NPN	0.45	10.0	0.002	0.000
Soy hulls	0.81	12.0	0.724	0.051					
Soybean meal, 44%	0.85	49.6	0.000	1.000					
Soybean meal, 48%	0.85	54.0	-0.093	1.111					
Soybeans, whole	1.00	37.7	0.582	0.699					
Urea, feed grade	0.00	281.0	-6.707	6.843					
Wheat	0.92	11.5	0.968	0.038					
Wheat bran	0.72	18.0	0.406	0.202					
Wheat mids	0.83	18.7	0.624	0.220					
Whey, liquid	0.89	13.0	0.873	0.076					

Source: Michigan Dairy Nutrition Manual 1990.

200

Feedstuffs	Dry matter (%)	Metabolizable energy (Mcal/lb)	Energy conversion factor	Crude protein (%)	Protein conversion factor	Maximum recommended dietary level (% of DM)	Cost of soybean meal ($/ton)					
							Value ($/ton [as-fed])					
							80	133	160	186	214	240
Blood meal	92	1.08	0.78	86.0	1.95	5	156.67	264.56	318.50	372.45	426.39	480.34
Brewers grains	91	1.09	0.79	27.2	0.62	30	45.67	79.42	96.30	113.17	130.05	146.93
Brewers grains	23	1.09	0.79	27.0	0.61	30	11.45	19.92	24.15	28.38	32.62	36.85
Broiler litter	89	1.09	0.79	24.5	0.56	35	39.75	69.48	84.35	99.22	114.08	128.95
Canola meal	90	1.25	0.90	36.5	0.83	14	65.00	109.80	132.20	154.59	176.99	199.39
Corn gluten feed	40	1.36	0.99	26.2	0.60	90	21.30	35.59	42.74	49.88	57.03	64.17
Corn gluten feed	92	1.36	0.99	26.2	0.60	50	48.99	81.86	98.30	114.73	131.17	147.60
Corn gluten meal	90	1.46	1.06	67.2	1.53	8	125.07	207.54	248.78	290.01	331.25	372.49
Cottonseed meal	90	1.31	0.95	41.0	0.93	10	74.34	124.66	149.82	174.98	200.14	225.30
Cottonseeds	93	1.58	1.14	23.9	0.54	20	48.82	79.13	94.28	109.44	124.20	139.75
Cull beans	90	1.38	1.00	25.3	0.58	25	46.58	77.63	93.15	108.68	124.59	139.73
Distillers grain	93	1.41	1.02	23.0	0.52	60	44.30	73.47	88.05	102.63	117.22	131.80
Feather meal	93	1.15	0.83	91.3	2.07	5	169.70	285.48	343.38	401.27	459.16	517.05
Fish meal	90	1.22	0.88	67.0	1.52	10	120.62	202.85	243.96	285.08	326.19	367.30
Linseed meal	90	1.28	0.93	39.8	0.90	15	71.61	120.45	144.88	169.30	193.72	218.14
Malt sprouts	94	1.17	0.85	28.1	0.64	30	50.40	86.42	104.43	122.44	140.45	158.46
Meat scraps	94	1.17	0.85	54.8	1.25	10	101.74	171.98	207.10	242.23	277.35	312.47
Meat/bone meal	93	1.09	0.79	50.4	1.15	10	90.88	154.80	186.76	218.71	250.67	282.63
Peas	89	1.43	1.04	25.3	0.58	25	46.88	77.58	92.94	108.29	123.64	138.99
Safflower meal	92	0.94	0.68	25.4	0.58	15	40.24	72.10	88.03	103.97	119.90	135.83
Soybean meal	89	1.38	1.00	44.0	1.00	12	80.10	133.50	160.20	186.90	213.60	240.30
Sunflower meal	93	1.09	0.77	49.8	1.13	10	89.35	152.51	184.08	215.66	247.27	278.82
Thin stillage	5	1.45	1.05	29.7	0.67	10	3.10	5.13	6.14	7.15	8.17	9.18
Urea	100	0.00	0.00	281.0	6.39	1	511.20	849.87	1022.05	1315.64	1507.23	1698.82
Whole soybeans	92	1.50	1.08	42.8	0.97	25	82.47	136.17	163.01	189.86	216.71	243.55

Calculations

Price of alternative protein source (feed x). Feed x = [(Price of SBM × crude protein conversion factor) + (energy conversion factor − 1) ÷ 1.48 × (price of corn (DMB)] × (dry matter of feed x/100).

yards of material entering a landfill. Waste pulpers reduce the size of food waste items such as a head of lettuce, a loaf of bread, a pork chop, etc. to produce a consistent size, uniformly distributed product of similar moisture content. Waste pulpers, regardless of the form or dry matter content of the food waste entering the pulper, produce a rather constant product containing 50 to 60% moisture.

Pulped food waste from the University residence hall cafeterias is collected three times daily following each meal, deposited in 30-gallon polyethylene containers, and stored overnight in refrigerated storage rooms. The following morning, food waste is delivered by truck to the beef cattle feed center where it is utilized in various research projects. Fresh pulped food waste, as referred to at ISU, consists of postconsumer plate scrapings containing uneaten food and waste paper products such as napkins, cups, etc. and preconsumer unserved prepared foods, both of which have been processed through a Hobart waste pulper. The first study evaluating human food waste as a feedstuff for ruminants involved the feeding of fresh pulped food waste (FPF) as a dietary ingredient in total mixed rations (TMR) to beef cows.

FPF waste replaced 50% of the corn silage and all of the supplemental protein in treatment diets of a corn silage and soybean meal and shelled corn-based control diet. Selected element and fractionate composition of the FPF fed in this study is shown in table 11.7. The dry matter (DM) for FPF is higher than the DM for plate scrapings reported by Flores et al. (1993) and the institutional garbage cited by Jurgens (1993), which were 34.87 ± 7.28% and 17.9%, respectively. The higher DM content reported in this study may be attributed to the inclusion of waste paper products such as napkins, milk cartons, and cardboard. Corrugated cardboard was used to clean the pulpers following food-waste processing at each meal. In addition, because of their design, waste pulpers can reduce the free water content of high-moisture foods.

Cellulose represented 95.7% of the acid detergent fiber (ADF) fraction suggesting that the fiber portion of FPF should be digestible by the ruminant. Acid detergent lignin (ADL) represented 0.60% of the DM and 2.95% of the ADF fraction of FPF. For comparison, NRC (1984) reports ADL values for corn silage of 7% and corn grain of 1%.

The analyzed crude protein (CP) percent for FPF in this study was similar to previous reports for FPF (Walker and Wertz 1994; Walker et al. 1997) and was in agreement with the CP percent cited for dehydrated edible restaurant waste (Myer et al. 1994). The percent ether extract (EE) determined for FPF is similar to values reported in the literature (Flores et al. 1993; Jurgens 1993; Kornegay et al. 1970). Ether extract values of this magnitude are not surprising since the typical American

Table 11.7. Mean (±SD) of selected element and selected fractionate composition (dry matter basis) of fresh pulped food waste fed during trials 1, 2, and 3

	Element composition											
Item	Ca (%)	P (%)	K (%)	Mg (%)	S (%)	Na (%)	Zn (ppm)	Mn (ppm)	CU (ppm)	Fe (ppm)	Co (ppm)	Al (ppm)
FPF[a]	0.84	0.49	0.42	0.08	0.28	0.56	36.98	11.49	4.73	124	1.13	530
SD	±0.32	±0.15	±0.06	±0.02	±0.04	±0.10	±6.63	±6.22	±3.21	±74	±0.27	±447
n	44.00	44.00	44.00	44.00	44.00	44.00	44.00	44.00	44.00	44	44.00	44
MTL[b]	2.00	1.00	3.00	0.40	0.40	10.00	300.00	1000.00	25.00	500	10.00	1000

	Fractionate composition							
Item	DM (%)	ADF (%)	Cellulose (%)	ADL (%)	ASH (%)	N (%)	CP (%)	EE (%)
FPF[a]	46.14	20.33	19.46	0.60	0.27	4.70	29.36	15.84
SD	±9.59	±6.79	±6.79	±1.12	±0.23	±1.11	±7.22	±3.25
n	143.00	140.00	140.00	94.00	106.00	144.00	144.00	144.00

[a]Fresh pulped food waste.
[b]NRC estimates for maximum tolerable level in livestock.

diet contains 40% of the daily calorie intake as fat (Kris-Etherton et al. 1988).

None of the 13 elements evaluated in FPF were found to exceed the maximum tolerable levels (MTL) recommended by the NRC (1984) for beef cattle diets. Fresh pulped food waste was found to contain an acceptable ratio of calcium:phosphorus of 1.95:1. Myer et al. (1994) reported dehydrated edible restaurant waste to contain 0.9% sodium (wet weight basis, moisture content equal to 7.9%) and expressed concern about potentially high levels of sodium in food waste recycled as a feedstuff. The mean sodium content ($0.56 \pm 0.10\%$) of FPF in the ISU study exceeds the sodium content of many traditional feedstuffs (Jurgens 1993) such as vegetative alfalfa (IFN = 2-00-181) 0.21%, corn grain (IFN = 4-02-931) 0.02%, and soybean meal (IFN = 5-04-612) 0.03%, but is substantially lower than some potential feedstuffs such as dehydrated bakery waste (IFN = 4-00-466) 1.24%, carrots (IFN = 4-01-145) 1.04%, and sugar beet molasses (IFN = 4-00-668) 1.48%. High elemental sodium might be expected in human food waste as the human diet often contains sodium levels higher than the recommended daily allowance (Kris-Etherton et al. 1988). However, cows fed diets containing FPF consumed similar amounts of a commercial mineral mixture containing 11% sodium chloride when the mineral mixture was offered free-choice.

The selected fractionate composition of the control diets (CTL) and diets containing fresh-pulped food waste (TRT) are reported in table 11.8. Percent ether extract was significantly higher in TRT than CTL in each trial, ranging from 1.36 to 2.44 times greater. Crude protein was higher ($P < 0.05$) for TRT than CTL during trial 2 (T2) and trial 3 (T3) but was not significantly different during trial 1 (T1). Acid detergent fiber and acid insoluble ash (AIA) were higher ($P < 0.05$) for CTL and TRT during T3 but were not significantly different during T1 and T2. The amount of FPF fed (mean = $45.4 \pm 6.3\%$) ranged from 37.0 to 52.3% of the TMR (table 11.9). Cows appeared to readily accept FPF as part of their TMR at the levels fed. No differences ($P > 0.05$) in cracked corn or average daily feed intake (ADFI) were observed between TRT and CTL. Due to replacing a portion of the forage with FPF, mean daily forage intakes of the cows in each trial were significantly lower for TRT than CTL. The lower FPF consumption during T3, period 2 (P2) was the direct result of knowingly feeding less FPF due to greater bone content of the food waste. Food waste containing bones was composted with ground paper and was not utilized as a feedstuff. During T3, P2 a substantial portion of meals were composed of foods containing bones that resulted in less FPF available for feeding.

Table 11.8. Mean (±SD) percent selected fractionate composition (dry matter basis) of control and treatment diets

Treatment	DM	ADF	EE	CP	CELL	ADL	AIA
				Trial 1[c]			
CTL[a]	59.9 ± 4.4	20.9 ± 1.5	2.8 ± 0.1	10.6 ± 0.7	16.2 ± 1.4	3.5 ± 0.8	1.2 ± 0.4
TRT[b]	50.2 ± 3.8*	18.01 ± 0.1	5.3 ± 0.5*	10.5 ± 0.7	14.8 ± 1.7	2.5 ± 1.2	0.7 ± 0.2
n	10	10	10	10	10	10	10
				Trial 2—Period 1[c]			
CTL	47.5 ± 19.3	23.7 ± 4.2	2.7 ± 0.2	11.2 ± 2.0	18.3 ± 2.7	3.9 ± 1.7	1.5 ± 0.6
TRT	46.3 ± 20.5	20.9 ± 3.2	6.6 ± 1.4*	13.4 ± 1.9*	16.9 ± 2.4	2.8 ± 1.3	1.2 ± 0.7
n	8	8	8	8	8	8	8
				Trial 2—Period 2[c]			
CTL	52.3 ± 8.4	19.6 ± 1.1	2.8 ± 0.2	10.7 ± 2.4	15.8 ± 1.3	2.9 ± 0.3	1.0 ± 0.3
TRT	52.5 ± 9.3	19.1 ± 1.3	6.5 ± 1.0*	12.1 ± 2.0*	15.4 ± 1.3	2.9 ± 2.07	0.8 ± 0.2
n	14	14	14	14	14	14	14
				Trial 3—Period 1[d]			
CTL	61.8 ± 4.4	28.8 ± 4.6	4.5 ± 1.0	17.9 ± 3.0	19.9 ± 2.4	7.0 ± 2.1	2.3 ± 1.5
TRT	50.2 ± 9.8*	21.3 ± 4.9*	10.0 ± 3.1*	18.9 ± 4.7*	15.1 ± 2.5	5.3 ± 2.5	0.9 ± 0.7*
n	8	8	8	8	8	8	8
				Trial 3—Period 2[d]			
CTL	61.4 ± 19.3	28.0 ± 2.4	5.8 ± 1.5	18.0 ± 4.4	20.6 ± 2.0	5.6 ± 1.6	1.8 ± 0.6
TRT	58.6 ± 10.4	20.9 ± 5.0*	7.9 ± 1.7*	20.2 ± 4.0*	16.0 ± 4.5	4.0 ± 1.3	0.8 ± 0.4*
n	14	14	14	14	14	14	14

[a]Control diets.
[b]Diets containing fresh pulped food waste.
[c]Forage component of diets consisted of corn silage.
[d]Forage component of diets consisted of soybean silage.
*Means within each period, within a column with different superscripts differ ($p < 0.05$).

Table 11.9. Mean (±SD) daily (wet weight basis) total mixed ration intake per cow by treatment (kg)

Treatment	Forage	FPF	Cracked corn	Soybean meal	ADFI	FPF (%)
			Trial 1[c]			
CTL[a]	17.4 ± 2.16*	—	4.0 ± 0.13	1.0	22.4 ± 1.95	
TRT[b]	8.7 ± 1.37	8.7 ± 1.12	4.0 ± 0.17	—	21.4 ± 1.78	40.6
			Trial 2—Period 1[c]			
CTL	15.9 ± 0.45*	—	1.0 ± 0.04	1.2	18.1 ± 0.54	
TRT	8.0 ± 0.67	8.5 ± 0.17	1.0 ± 0.03	—	17.5 ± 0.66	48.4
			Trial 2—Period 2[c]			
CTL	17.6 ± 1.18*	—	1.6 ± 0.13	0.6	19.8 ± 2.01	
TRT	8.7 ± 2.04	9.8 ± 0.95	1.7 ± 0.15	—	20.2 ± 1.26	48.7
			Trial 3—Period 1[d]			
CTL	20.2 ± 0.37*	—	1.5 ± 0.01	—	21.7 ± 0.92	
TRT	8.9 ± 0.09	11.4 ± 0.05	1.5 ± 0.09	—	21.8 ± 0.87	52.3
			Trial 3—Period 2[d]			
CTL	19.0 ± 4.25*	—	2.4 ± 0.02	—	21.4 ± 2.45	
TRT	11.3 ± 1.18	8.1 ± 3.41	2.5 ± 0.19	—	21.9 ± 3.65	37.0

[a]Control diets.
[b]Diets containing fresh pulped food waste.
[c]Forage component of diets consisted of corn silage.
[d]Forage component of diets consisted of soybean silage.
*Means within each period, within a column with different superscripts differ ($p < 0.05$).

Similar cow weight changes and body condition score changes (except T1) were observed (table 11.10) in the ISU study. Even though cows fed TRT in T1 lost body condition while cows fed CTL gained condition, the condition scores for cows on both diets remained in the acceptable 5 category throughout the trial. Body condition score changes fluctuated little throughout T1, T2, or T3 for either control or treatment cows suggesting that dietary management was successful in achieving similar condition score changes in an optimum condition score range. No differences (P > 0.05) in body weight changes between treatment and control cows were observed throughout the study. Body weight changes observed reflect normal expected fluctuations for cows during the last trimester and first 60 to 90 days postpartum (Wiltbank et al. 1964).

Statistical differences in the measures of cow production performance were observed (table 11.11). Calf age at trial termination, ending weights, and average daily gains (ADG) were higher (P < 0.05) for calves nursing cows fed food waste compared to controls. Differences in calf ending weights and ADG between cows fed CTL and TRT due to significant differences in calf age should have been accounted for since calf age was used as a covariate in the calf ending weight and ADG analyses. No significant differences in milk production, milk fat, or milk protein were observed. Records of creep feed intake by calves were not kept and differences in feed intake could have contributed to individual calf performance. It is possible that calves nursing cows fed food waste consumed more creep feed or ate more feed at the bunk with their dams, which could have contributed to the significantly higher calf ending weights and average daily gains. Significant differences (P < 0.05) in cow conception rates were observed between years and in T2 between cows fed CTL and TRT. Inclusion of fresh-pulped food in the diets of the beef cows could have had an effect on conception rates but the data collected do not offer an explanation for this observation.

Table 11.12 reports the relative economic value of fresh-pulped food waste. The values reported are based on the data of the ISU study only and reflect FPF value as a substitute for soybean meal and as a partial replacement for corn silage or soybean silage in the beef cow's diet.

On a wet weight basis when substituted for soybean meal and corn silage, FPF was calculated to have a comparative, replacement value ranging from 2.93 to 4.364:kg (1.33 to 1.984:lb). The relative value of FPF varies directly with the amount and cost per unit of the feedstuff(s) FPF replaces in the diet. Trials 1 and 2 found FPF, on a replacement basis, to be worth 1.44 to 2.39 times the cost of corn silage based on the market prices of the feedstuffs quoted. As a replacement feedstuff for

Table 11.10. Body weight (kg) and condition score absolute values and changes (mean ±SD) of cows fed CTL[a] and TRT[b]

Treatment	Starting weight	Ending weight	Weight change	Starting condition score	Ending condition score	Condition score change
			Trial 1			
CTL	715 ± 40	712 ± 57	−3 ± 19	5.0 ± 0.1	5.3 ± 0.2	0.3 ± 0.2
TRT	711 ± 40	685 ± 47	−26 ± 10	5.3 ± 0.2	5.1 ± 0.1	−0.2 ± 0.2[d]
			Trial 2—Period 1			
CTL	710 ± 82	732 ± 93	22 ± 29	5.8 ± 0.8	5.5 ± 0.6	−0.3 ± 0.6
TRT	705 ± 73	720 ± 73	15 ± 2	5.8 ± 0.8	5.5 ± 0.7	−0.3 ± 0.5
			Trial 2—Period 2			
CTL	732 ± 93	723 ± 142	−9 ± 30	5.5 ± 0.6	6.0 ± 1.0	0.5 ± 0.4
TRT	720 ± 73	701 ± 77	−19 ± 25	5.5 ± 0.7	6.0 ± 1.0	0.5 ± 0.3
			Trial 2—Period 1 and Period 2 Combined			
CTL	710 ± 82	723 ± 142	13 ± 50	5.8 ± 0.8	6.0 ± 1.0	0.2 ± 0.8
TRT	705 ± 86	701 ± 77	−4 ± 60	5.8 ± 0.8	6.0 ± 1.0	0.2 ± 0.7
			Trial 3—Period 1[c]			
CTL	755 ± 92	783 ± 69	27 ± 26	5.5 ± 0.5	5.7 ± 0.6	0.2 ± 0.4
TRT	755 ± 71	767 ± 63	12 ± 19	5.5 ± 0.6	5.7 ± 0.6	0.2 ± 0.3
			Trial 3—Period 2			
Cows						
CTL	783 ± 69	747 ± 66	−36 ± 33	5.7 ± 0.6	5.2 ± 0.5	−0.5 ± 0.5
TRT	767 ± 63	752 ± 81	−15 ± 46	5.7 ± 0.6	5.2 ± 0.7	−0.4 ± 0.5
Heifers						
CTL	648 ± 60	681 ± 72	33 ± 28	5.4 ± 0.4	5.4 ± 0.4	0.0 ± 0.7
TRT	646 ± 50	659 ± 55	13 ± 38	5.4 ± 0.4	5.2 ± 0.5	−0.1 ± 0.4
			Trial 3—Period 1 and Period 2 Combined[c]			
CTL	769 ± 14	765 ± 18	−4 ± 31.5	5.6 ± 0.1	5.5 ± 0.3	−0.1 ± 0.4
TRT	761 ± 6	760 ± 7.5	−1 ± 13.5	5.6 ± 0.1	5.5 ± 0.3	−0.1 ± 0.3

[a]Control diets.
[b]Diets containing fresh pulped food waste.
[c]Means include data of multiparous cows only. Data from first parity cows is not included.
[d]Means within each period, within a column with different superscripts differ ($p<0.05$).

Table 11.11. Production performance (mean ±SD) of cows fed CTL[a] and TRT[b]

Treatment	Ending calf weight (kg)	Calf ADG (kg)	Ending calf age (d)	Calf crop saved-birth (%)	Calf crop weaned (%)	Cow conception rate (%)	Estimated milk production (kg)	Milk fat (%)	Milk protein (%)
Trial 1									
CTL	121.4 ± 3.1	1.07 ± 1.05	65.0 ± 4.6	94.0	94.0	90.0			
TRT	130.5 ± 6.2	1.17 ± 0.13	73.0 ± 6.2	97.0	91.0	84.0			
Trial 2									
CTL	106.4 ± 29.2	0.97 ± 0.24	60.0 ± 5.5	94.0[c]	89.0	94.0[d]	10.9 ± 5.5		
TRT	116.4 ± 34.5	1.02 ± 0.26	65.0 ± 11.0	86.0[c]	83.0	70.0[d]	10.9 ± 5.5		
Trial 3									
CTL	113.5 ± 28.3	1.11 ± 0.30	58.8 ± 16.1	100.0	95.0	76.0	8.8 ± 2.7	3.81 ± 0.93	
TRT	118.6 ± 23.5	1.16 ± 0.24	59.4 ± 13.7	95.0	94.0	76.0	6.6 ± 2.3	3.45 ± 0.66	3.30 ± 0.27
Combined									
CTL	113.8 ± 6.1*	1.05 ± 0.06*	61.3 ± 2.68*	96.0 ± 2.8	92.7 ± 2.6	87.3 ± 6.8			
TRT	121.8 ± 6.2*	1.12 ± 0.07*	65.8 ± 5.58*	92.7 ± 4.8	89.3 ± 4.6	76.7 ± 5.7			

[a]Control diets.
[b]Diets containing fresh pulped food waste.
[c]Within a trial, means within a column tend to differ ($p < 0.1$).
[d]Within each trial, means within a column differ ($p < 0.05$).
*Means for Trial 1, 2 and 3 combined, within a column differ ($p < 0.05$).

209

Table 11.12. Relative economic value of fresh pulped food waste[a]

Treatment	Total feed cost·cow^{-1}·d^{-1} ($)[b]	FPF fed·cow^{-1}·d^{-1} (kg)	Calculated value for FPF ($·kg^{-1})
		Trial 1	
CTL[c]	0.99	—	
FPF[d,e]	0.61	8.7	4.36
		Trial 2—Period 1	
CTL	0.66	—	
FPF[e]	0.26	8.5	4.73
		Trial 2—Period 2	
CTL	0.65	—	
FPF[e]	0.36	9.9	2.93
		Trial 3—Period 1	
CTL	1.32	—	
FPF[f]	0.67	11.4	5.70
		Trial 3—Period 2	
CTL	1.35	—	
FPF[f]	0.92	8.1	5.32

[a]All calculations based on wet weight basis.
[b]Cost of all dietary ingredients except food waste.
[c]Control diet.
[d]Diets containing fresh pulped food waste.
[e]Calculated value for FPF is based on corn silage values at 0.9¢: lb (1.98¢: kg), shelled corn valued at 5.0¢: lb (11.0¢: kg), and soybean meal valued at 9.5¢: lb (20.9¢: kg).
[f]Calculated value for FPF is based on soybean silage values at 2.6¢: lb (5.72¢: kg), soybeans at 11.7¢: lb (25.7¢: kg), and shelled corn at 5.0¢: lb (11.0¢: kg). Assumes soybeans yield 30 bushels of grain: acre or 4 tons of soybean silage: acre.

soybean silage (T3), FPF was calculated to have a comparative, replacement value of 5.32 to 5.704:kg (2.42 to 2.594:lb).

The data of this study reflect that FPF is an acceptable dietary ingredient for beef cows and that satisfactory intakes can be achieved over sustained periods of time when FPF is included at rates up to 52% of the total mixed ration. The inclusion of fresh pulped food waste in beef cows' diets does not appear to be detrimental to either the cows' health or to the health of the calf nursing a cow consuming FPF. Fresh pulped food waste appears to have value as both an energy substitute and as a protein replacement feedstuff. Economically, the relative replacement value of FPF suggests that including food waste in the diets of beef cows could save 31.9 to 60.6% per year in traditional feed cost. This savings assumes no purchase nor delivery charges for FPF.

In the fall of 1997 and subsequent to conducting these studies, the FDA adopted the Mammalian Protein-Ruminant Feed Ban which is aimed at preventing bovine spongiform encephalopathy. This ban excludes feeding ruminants food waste containing animal protein that has not been heat processed. The ban does, however, include the following exemption "inspected meat products which have been cooked and offered for human food and further heat processed for feed (such as plate waste and used cellulosic food casing)." The phrase "further heat processed" may include cooking at 212°F for 30 minutes, dehydration, and extrusion. Consequently, it is now illegal to feed FPF containing animal protein to cattle and sheep. However, vegetarian food waste may be fed without heat processing and similar performance as observed in the previous study could be expected. While it is no longer legal to feed food waste that has not been appropriately heat processed, the data presented in this study does show the potential nutritional and economic benefits of including food waste as part of the diet of beef cows.

Extruded Food Waste and Control of Potential Pathogens

There is concern that food waste may contain biological and chemical contaminants that may limit its reutilization as animal feed. Disease-causing microorganisms (pathogens) are of particular concern, especially in waste reutilized as animal feed. These pathogens could potentially be transferred to animals through ingested feed and cause infectious disease that might then be transferred to human consumers. In order to reduce risks associated with reutilization of waste as animal feed, some type of pretreatment is usually advisable to reduce or eliminate potentially pathogenic microorganisms prior to ingestion by animals. Pretreatments include boiling, chemical additives, ensiling, composting, or otherwise heating food waste to reduce or eliminate pathogen contamination depending on whether or not the food waste contains animal protein (Federal Register 1997.21; Kroyer 1995; Kelley et al. 1994; Kelley et al. 1995; Troeger et al. 1983).

Extrusion is a technique used in food processing to produce a typically light-texture, low-density food product (e.g., breakfast cereals, dry snack foods, dry pet foods, etc.). Heat and pressure are developed by passing previously mixed animal feed or human food ingredients through a barrel by means of a screw die with increasing restrictions, ultimately discharging the product into the atmosphere. Expansion of the product occurs with a sudden decrease in pressure (from approximately 2,700 PSI) when the material is discharged through the die into the atmosphere. The amount of expansion depends upon several factors, including the starch content of the material, moisture content (typically 20 to 30%), temperature, and pressure. The dry-extrusion method utilizes friction as the sole source of heat accompanied by pressure and attrition. The extrusion process takes less than 30 seconds to cook and dehydrate the product with typical product internal temperatures rang-ing from 140 to 160°C. A temperature of 121°C must be maintained for at least 15 min at approximately 100 PSI to ensure destruction of both vegetative bacterial cells and spores (sterilization). This suggests that any disinfection that occurs during extrusion may be due to the combined effects of heat, dehydration, and cell rupturing occurring during the abrupt change in pressure.

Consequently, a study was designed by investigators at ISU (Walker and Kelley 1997) to determine and compare the relative concentrations of pathogen indicator bacteria and other bacteria in raw food waste, food waste-amended animal feed prior to and following extrusion, and commercial swine feed. Bacterial groups and genera isolated and enu-merated were total and fecal coliform, enterococci, staphylococci, and heterotrophic bacteria. Selected samples of food waste and animal feed were also analyzed for nonspecific anaerobic/facultative bacteria. This information was used to determine the ability of a single-screw, dry-extrusion process to reduce concentrations of potentially pathogenic bacteria from feed prior to ingestion by animals. Bacterial concentration reduction in extruded animal feed should result in a product that has a decreased risk of transferring disease agents to animals, including humans.

Raw food waste was collected from student cafeterias at ISU, pulped, mixed with other feed ingredients, and dry-extruded to produce animal feed. Food waste used in this study was composed of postconsumer plate scrapings containing uneaten food and waste paper products such as napkins, cups, etc., and preconsumer unserved prepared foods. Cafete-ria dishroom staff scraped food waste into continuous-flow recirculating water troughs, which utilized recirculated dishwater from automated dishwashers. Waste troughs then transported the food waste to Hobart waste pulpers (Hobart Co., Troy, OH).

Pulped food waste was mixed with soybean hulls and ground corn at a ratio of 40:55:5 (wet weight), respectively. Feed was mixed in quanti-ties of approximately 180 kg, using a horizontal ribbon mixer with a capacity of 225 kg prior to extrusion. The feed mixture was then extruded at temperatures ranging from 110 to 135°C utilizing an Insta-Pro®, model 600 JR extruder (InstaPro International, Des Moines, IA). The extruder barrel consisted of one single screw, four 9.53 cm steam locks alternating with three double screws, and the head assembly with one single 0.79 cm diameter die, absent a cutter. The extrusion process was replicated 16 times over a 6-month period. Preextrusion feed moisture content ranged from 31 to 41% with a mean of 37 ± 3%. Postextrusion moisture content ranged from 12 to 37%, with a mean of 31 ± 5%.

Postextrusion feed samples were collected exiting the die, allowing the extruded product to fall into sterile, 480 ml (16 oz), glass jars that were then sealed and delivered for bacterial analyses. Samples of raw food waste, preextrusion animal feed, and commercial swine feed were also collected aseptically and transferred into sterile glass jars prior to delivery for bacterial analyses. The weight of samples collected was approximately 200 to 300 g. One to three pre- and postextrusion samples were collected from each extruded feed batch. One 10 g subsample of each sample collected was analyzed for total and fecal coliform, enterococci, staphylococci, and heterotrophic bacteria using standard culturing methods. Selected samples were also analyzed for nonspecific anaerobic/facultative bacteria.

Concentrations of total coliform in raw food waste ranged from below the detection limit to 5.80×10^2 cfu/g dry weight (mean = 2.57×10^2). Two of six raw food-waste samples analyzed were found to have total coliform concentrations below the detection limit. Fecal coliform and enterococci concentrations were below the detection limit in all raw food-waste samples analyzed. Concentration of Staphylococci in raw food waste ranged from the detection limit to 2.13×10^4 cfu/g dry weight (mean = 8.68×10^3). Four of eight raw food waste samples analyzed were found to have Staphylococci concentrations below the detection limit. Concentrations of heterotrophic bacteria recovered from raw food waste ranged from 1.84×10^3 to 8.95×10^3 cfu/g dry weight (mean = 4.66×10^3). A relatively small number of raw food waste samples were analyzed ($N = 6$ to 8), due to the primary focus of this study on changes in bacterial concentrations in animal feed from pre- to postextrusion.

Concentrations of total coliforms in preextrusion animal feed ranged from below the detection limit to 1.48×10^4 cfu/g dry weight (mean = 4.02×10^3). Concentrations of total coliform below the detection limit were found in only two of 36 samples analyzed. Fecal coliform concentrations in preextrusion animal feed ranged from below the detection limit to 1.60×10^4 cfu/g dry weight (mean = 4.37×10^2). Fecal coliform concentrations were below the detection limit in 15 of 36 samples analyzed. Enterococci concentrations recovered from preextrusion animal feed ranged from the detection limit to 7.72×10^3 cfu/g dry weight (mean = 1.93×10^3). Staphylococci concentrations in preextrusion animal feed ranged from 1.90×10 to 8.35×10^3 cfu/g dry weight (mean = 2.16×10^3). Heterotrophic bacteria concentrations in preextrusion animal feed ranged from 4.53×10^3 to 3.19×10^4 cfu/g dry weight (mean = 1.73×10^4).

Total and fecal coliform, enterococci, and Staphylococci concentrations in postextrusion animal feed were below the detection limit in all samples analyzed (total of 41). Concentrations of heterotrophic bacteria ranged from below the detection limit to 2.44×10^4 cfu/g dry weight (mean = 4.96×10^3). Approximately half (21 of 41) of the samples analyzed were below the detection limit for heterotrophic bacteria. Evidence of nonspecific anaerobic/facultative bacteria was found in 10 of 20 postextrusion samples analyzed. Concentrations of anaerobic/facultative bacteria in postextrusion animal feed ranged from below the detection limit to 3.85×10^4 cfu/g dry weight (mean = 1.34×10^4).

Concentrations of total coliform in commercially-available swine feed ranged from below the detection limit to 2.60×10^4 cfu/g dry weight (mean = 7.86×10^3). Only two of 29 swine feed samples analyzed were found to be below the detection limit for total coliforms. Fecal coliform concentrations in swine feed ranged from below the detection limit to 7.58×10^3 cfu/g dry weight (mean = 2.53×10^3). Nine of 29 swine feed samples were found to be below the detection limit for fecal coliform concentrations. Enterococci concentrations in swine feed samples analyzed ranged from the detection limit to 4.49×10^3 cfu/g dry weight (mean 2.53×10^3). Fourteen of 29 swine feed samples analyzed were found to be below the detection limit for enterococci concentrations. Staphylococci concentrations in swine feed ranged from 9.18×10^2 to 1.11×10^4 cfu/g dry weight (mean = 3.92×10^3). Concentrations of heterotrophic bacteria in swine feed samples ranged from 5.51×10^3 to 5.63×10^4 cfu/g dry weight (mean = 2.34×10^4). While only six commercial swine feed samples were analyzed for nonspecific anaerobic/facultative bacteria, all six samples analyzed were found to contain concentrations of anaerobic/facultative bacteria exceeding 9.44×10^3 cfu/g dry weight (mean = 6.17×10^4). This is more than four times the mean concentration of anaerobic/facultative bacteria recovered from postextrusion feed samples (1.34×10^4 cfu/g dry weight).

Results of this study indicated that single-screw, dry-extrusion processing treatment substantially reduced concentrations of indicator bacteria. Concentrations of total and fecal coliform, enterococci, staphylococci, and heterotrophic bacteria concentrations decreased substantially from highest initial preextrusion concentrations of 2 to 4 $\log_{10}$ to final concentrations below the detection limit in a majority of samples tested following extrusion. Recovery of heterotrophic and nonspecific anaerobic/facultative bacteria in concentrations above the detection limit in postextrusion feed samples indicated that the extrusion process did not consistently sterilize animal feed (table 11.13). Increased extrusion temperatures generally resulted in decreased bacterial concentrations in postextrusion samples analyzed.

Table 11.13. Concentrations (cfu/g dry weight) of microbial groups and genera isolated from raw food waste, food waste-supplemented animal feed before and after extrusion, and commercial swine feed samples (mean ± SD)

	Total coliform	Fecal coliform	Enterococci	Staphylococci	Heterotrophic bacteria	Anaerobic/facultative bacteria
Raw food waste						
Mean ± 1 SD	2.57×10^2 $\pm\, 2.15 \times 10^2$	$2.00 \times 10^{0b} \pm 0$	$2.00 \times 10^0 \pm 0$	8.68×10^3 $\pm\, 9.11 \times 10^3$	4.66×10^3 $\pm\, 2.58 \times 10^3$	NA[d]
N[a]	4	6	8	4	6	NA
MDL[b]	2	6	8	4	0	NA
Total[c]	6	6	8	8	6	NA
Animal feed (pre-extrusion)						
Mean ± 1 SD	4.02×10^3 $\pm\, 4.08 \times 10^3$	4.37×10^3 $\pm\, 4.87 \times 10^3$	1.93×10^3 $\pm\, 2.07 \times 10^3$	2.16×10^3 $\pm\, 2.39 \times 10^3$	1.73×10^4 $\pm\, 8.12 \times 10^3$	NA
N	34	21	19	36	36	NA
MDL	2	15	17	0	0	NA
Total	36	36	36	36	36	NA
Animal feed (post-extrusion)						
Mean ± 1 SD	$2.00 \times 10^0 \pm 0$	$2.00 \times 10^0 \pm 0$	$2.00 \times 10^0 \pm 0$	$2.00 \times 10^3 \pm 0$	4.96×10^3 $\pm\, 6.94 \times 10^3$	1.34×10^4 $\pm\, 1.11^c + 4$
N	41	41	41	41	20	10
MDL	41	1	41	41	21	10
Total	41	41	41	41	41	20
Swine feed						
Mean ± 1 SD	7.68×10^3 $\pm\, 5.68 \times 10^3$	2.53×10^3 $\pm\, 2.62 \times 10^3$	2.33×10^3 $\pm\, 2.11 \times 10^3$	3.92×10^3 $\pm\, 3.72 \times 10^3$	2.34×10^4 $\pm\, 2.00 \times 10^4$	6.17×10^4 $\pm\, 4.00 \times 10^3$
N	27	20	15	29	29	6
MDL	2	9	14	0	0	0
Total	29	29	29	29	29	6

[a] (N) = Number of samples subjected to statistical analysis (mean and SD determination).
[b] (MDL) = Number of samples below minimum detection limits (MDL = 2.00×10^0 ± cfu/g dry weight).
[c] (Total) = Total number of samples analyzed.
[d] (NA) = Not applicable (analyses not performed).

Substantially higher concentrations of heterotrophic and nonspecific anaerobic/facultative bacteria were recovered from commercial swine feed samples than postextrusion animal feed samples (table 11.13). It is difficult to compare postextrusion animal feed samples to commercial swine feed because of differences in handling techniques, but it appeared that animals would ingest higher concentrations of bacteria in commercial feed samples, compared to postextrusion animal feed samples analyzed. However, there are many environmental sources of post-production bacterial feed contamination, including feed containers, feeding equipment, air, water, soil, humans, and other animals.

Soybean hulls and ground corn often contributed more to bacterial concentrations recovered from preextrusion animal feed than food waste (table 11.13). This may be due to bacterial contamination during processing and handling of the soybean hulls and ground corn. Relatively low bacterial concentrations recovered from food waste may be due to the disinfecting action of food cooking temperatures together with the addition of chemicals such as Quaternary Ammonium Compounds (Quats), commonly used as surface sanitizers in foodservice. Recirculated wastewater from dishwashers that contain automated sanitizer dispensers may have contained active sanitizing agents that had a residual disinfectant effect in the food waste. Since food waste was amended with other feed ingredients and extruded within 24 hours after collection, the added sanitizer still may have been active as a disinfectant.

The single-screw, dry-extrusion process substantially decreased bacterial concentrations and therefore the potential for transmission of infectious disease through feed ingredients, including food waste, to animals. Survival of heterotrophic bacteria following the extrusion process may be of some concern, since some of these bacteria may be pathogenic. However, the elimination of pathogen indicators such as coliform during the extrusion process makes this prospect unlikely due to increased survival of coliform compared to most other pathogens. Survival of nonspecific anaerobic/facultative bacteria may be of greater concern, due to the possibility of survival and subsequent germination of heat-resistant spores and transfer of infectious disease (e.g., clostridia). Further research to distinguish obligate anaerobes from facultative bacteria in postextrusion feed samples should help to determine associated risks. These concerns merit further investigation to improve the extrusion process to reduce bacterial contamination as much as possible, thereby reducing the associated risk of infectious disease transmission.

This research initiative is being continued. These investigations are utilizing a Model 2000 Insta-Pro® extruder. The model 600 is no longer used in these studies. The larger unit with greater capacity has greater capability for extruding waste materials. We have reconfigured the steam lock, and screw combinations are consistently operating the extruder at temperatures ranging from 295°F to 310°F. The researchers speculate that the larger the single-screw extruder, the greater the capacity for successfully processing high-moisture food waste. Current study is evaluating the Model 2000 extruder's capacity for disinfecting and sterilizing food-waste amended feeds.

Palatability and Digestibility of Extruded Food Waste

Table 11.14 shows the average fiber, protein, ether extract, and dry matter composition of all fresh pulped food-waste samples collected at ISU from the fall of 1993 through the spring of 1996. For reference, the dry matter of the food waste collected parallels the dry matter content of corn silage that normally ranges between 30 and 50%. Relative to the dry matter content of food waste or garbage previously reported (Flores et al. 1993; USEPA 1997; Kornegay et al. 1965; Kornegay et al. 1968), the food waste produced at ISU is considerably drier comparing 46.14 ± 9.59% to a range of 16.0 to 33.52%. The crude protein value observed in the ISU food waste (29.36 ± 7.22%) is considerably higher than previously reported values that have ranged between 14.6 and 17.6%. The ether extract (estimated crude fat) percent (15.84 ± 3.25%) of food waste collected at ISU is well within the ranges (14.7 to 32.0%) previously reported. Other investigators have not reported acid detergent fiber, cellulose, nor lignin values for food waste. While the ADF values for food waste are relatively high compared to the values reported in the NRC tables for grains, (corn, IFN 4-02-931, = 4.3%; oats, IFN 4-03-309, = 16.0%; wheat, IFN 4-04-211, = 4.13%) they are low compared to forages (corn silage, IFN 3-28-250, = 28%; fescue, IFN 1-10-871, = 39%; alfalfa, IFN 1-00-063, = 35%). Analyses by ISU investigators also observed the cellulose component of ADF to be relatively high and the lignin portion to be correspondingly low. Cellulose has been found to comprise 95.7% of ADF while the lignin makes up only 3.0%. The fiber fractionates of the ISU food waste suggest that ADF of food waste should be highly digestible in the rumen of sheep and cattle, that food waste has greater potential as a feedstuff for ruminants than nonruminants, and its potential as a feedstuff for nonruminants is greater for breeding stock than it is for growing-finishing animals.

Researchers at ISU have analyzed food waste for the concentration of several selected elements. Food waste (table 11.15) has a higher content of calcium than soybean meal, IFN 5-04-600, (0.83 ± 0.32% v. 0.29

Table 11.14. Mean (±SD) percent selected fractionate composition of pre-extruded pulped food waste (dry matter basis)

Item	DM	ADF	CELL	ADL	ASH	N	CP	EE
Mean	46.14	20.33	19.46	0.60	0.27	4.70	29.36	15.84
SD	9.59	6.79	6.79	1.12	0.23	1.11	7.22	3.25
N	143.00	140.00	140.00	94.00	106.00	144.00	144.00	144.00

Table 11.15. Mean (±SD) selected element composition of pre-extruded pulped food waste (dry matter basis)

Item	Ca	P	K	Mg	S	Na	Zn	Mn	Cu	Fe	Co	Al
	%	%	%	%	%	%	ppm	ppm	ppm	ppm	ppm	ppm
Mean	0.83	0.50	0.36	0.08	0.25	0.48	37.96	13.33	6.11	145	1.00	22
SD	0.32	0.17	0.06	0.02	0.05	0.11	7.46	8.67	4.17	105	0.00	730
n	27.00	27.00	27.00	27.00	27.00	27.00	27.00	27.00	27.00	27	27.00	27
MTL[1]	2.00	1.00	3.00	0.40	0.40	10.00	300.00	1000.00	25.00	500	10.00	1000

[1]MTL = Maximum Tolerable Limit from NRC (1980).

− 0.33%), a lower content of phosphorus (0.50 ± 17% v. 0.68 − 0.71%), lower contents of magnesium (0.08 ± 0.02% v. 2.0%) and potassium (0.36 ± 0.06 v. 0.29%) and a substantially higher sodium (0.48 ± 0.11 v. 0.03%) value, respectively (NRC 1985). Even at 0.48 ± 0.11% the sodium content of food waste does present a problem to the nutritionist when balancing diets for livestock. None of the element contents evaluated for food waste were found to exceed the NRC maximum tolerable limits (MTL) for livestock (NRC 1980).

A study was conducted at ISU to investigate the use of extrusion technology for producing a feedstuff from food waste, soybean hulls, and rolled corn. In addition, these researchers evaluated the efficacy of the extruded mixture as a feedstuff for ruminants through the determination of the selected fractionate composition of an extruded material containing pulped food waste, soybean hulls, and rolled shelled corn, and the conduct of a digestibility trial utilizing sheep to estimate the digestible energy and the digestibility coefficients for dry matter intake, protein, ether extract, acid detergent fiber, and cellulose. The objective of this study was to evaluate the potential for extruding food waste. It should be noted that unprocessed garbage does not lend itself to passing through the extruder. The pulping process, which is a form of wet grinding, puts food waste in an ideal form for extruding, except for its relatively high moisture content. The ideal moisture content of material to be extruded ranges from 20 to 30%. Initially these investigators tried several combinations of screw types, steam lock size and type, and size of exiting die in attempts to extrude raw pulped food waste. This material, without alteration, was too wet to extrude. Consequently, several attempts were made to pass pulped food waste through a continuous horizontal press. Horizontal presses are typically used to extract oil from seeds such as soybeans. Regardless of adjustments made to the press, the press would not process the pulped food waste. The intention was to press 50 to 60% of the water from the pulped food waste prior to extruding the food waste. Therefore, pulped food waste was blended with several other raw materials, prior to extruding, that were capable of absorbing moisture. Blended combinations included pulped food waste and ground paper, pulped food waste and ground shelled corn, and pulped food waste and soybean hulls.

Originally these researchers planned to produce a pellet as their end product. In order to produce a pellet of desirable texture and consistency, ground shelled corn was added to pulped food waste to absorb moisture and provide a source of soluble starch. Corn and pulped food waste were extruded in the following wet weight ratios, 75:25 and 77:23, respectively, which consistently produced a blended material prior to extrusion containing 24 to 28% moisture. One hundred thirty-six (136) kg of the 77:23 product was extruded and 1,091 kg of the 75:25 product was produced in several batches. Generally, they found these two products to be low in ADF (as expected, because shelled corn contains only 3% ADF) and similar to corn in crude protein. The moisture content of material exiting the extruder was dry, 89.31 and 90.00 ± 3.03% for the 77:23 and 75:25 products, respectively. This pellet was allowed to cool by scattering on a concrete floor for 24 hours.

Following drying, the pellet was offered to sheep (mature ewes) who nibbled but generally refused to eat the pellet. Approximately 955 kg of the 75:25 product was stored for 6 months in six polyethylene barrels. This product visually appeared to store very well with a notable absence of mold. Following storage the product was fed to 60 mature beef cows at the rate of 2.27 kg:head:day for 7 days as part of their total mixed ration. None of the pellets were refused as part of the total mixed ration.

To reduce the cost of the extruded product and to utilize more food waste, lower cost raw materials were blended with pulped food waste to replace some of the corn. After extruding food waste with several other feedstuffs, soybean hulls were utilized as the primary coextrusion material because of its dry nature, highly digestible fiber content, and low relative cost. Soybean hulls typically range in price from $35 to $70 per ton in the Midwest. As a result, combining food waste with soybean hulls and

Table 11.16. Mean (±SD) fractionate composition of ingredients prior to extrusion (dry matter basis)

Item	DM	ADF	CELL	ADL	ASH	N	CP	EE
			Pulped Food Waste					
Mean	43.66	14.55	9.73	4.49	0.33	4.50	28.12	7.15
SD	7.71	7.89	8.97	3.18	0.10	1.17	7.33	3.91
n	12.00	12.00	12.00	12.00	12.00	12.00	12.00	12.00
			Rolled Shelled Corn (IFN = 4-02-931)[1]					
	88.00	4.30	2.40	0.40	1.50	1.66	10.10	4.20
			Soybean Hulls (IFN = 1-04-560)[1]					
	90.30	49.00	46.10	50.00	4.90	1.98	9.00	2.10

[1]NRC 1984.

Table 11.17. Mean (±SD) percent selected fractionate composition of mixed ingredients[1] (dry matter basis)

Item	DM	ADF	CELL	ADL	ASH	N	CP	EE
			Pre-Extrusion					
Mean	74.65	30.76	25.17	4.45	1.14	2.48	15.49	4.86
SD	5.36	8.94	9.62	2.82	1.34	0.37	2.33	0.50
n	8.00	8.00	8.00	8.00	8.00	8.00	8.00	8.00
			Post-Extrusion					
Mean	84.23	33.47	28.94	3.66	0.87	2.62	16.39	4.41
SD	2.00	5.15	6.90	1.79	0.21	0.25	1.56	0.51
n	14.00	14.00	14.00	14.00	14.00	14.00	14.00	14.00

[1]Mixed ingredients include pulped food waste, soybean hulls, and rolled shelled corn in a 40:55:5 ratio, wet weight basis.

rolled shelled corn proved feasible. The most successful combination extruded was FW:SBH:RC in a 40:55:5 ratio (wet weight basis).

Table 11.16 shows the dry matter, fiber, protein, and ether extract values for pulped food waste, soybean hulls, and rolled shelled corn prior to extruding. The food waste fractionates represent values determined in the scientist's laboratory. The soybean hull and corn values represent NRC estimates. Table 11.17 provides the preextrusion and postextrusion dry matter, fiber, ether extract, and crude protein values for the mixed combination of FW:SBH:RC. Approximately 10 percentage points of moisture were lost during extrusion. Accordingly, the extruded product produced was able to be stored for an extended period without spoiling. Little change in fractionate composition was observed between preextruded and postextruded food-waste amended feed.

Table 11.18 shows the selected element composition of extruded food-waste amended feed (FW:SBH:RC) and contrasts the values to the

Table 11.18. Mean selected element composition of extruded food waste amended feed (EFWM) and daily requirements for specific livestock (dry matter basis)

Item	Ca (%)	P (%)	K (%)	Mg (%)	S (%)	Na (%)	Zn (ppm)	Mn (ppm)	Cu (ppm)	Fe (ppm)	Co (ppm)	Al (ppm)
					Element Composition							
EFWM[1]	0.46	0.22	1.23	0.23	0.16	0.11	42.14	17.00	7.71	378.4	1.00	141.4
SD	0.02	0.02	0.04	0.01	0.01	0.02	3.72	1.56	0.88	39.49	0.00	50.21
n	14	14	14	14	14	14	14	14	14	14	14	14
					Dietary Requirement For Gestating Females							
Sows	0.75	0.60	0.20	0.04	0.10	0.15	50	10	5	80	0.10	
Cows	0.26	0.22	0.65	0.10		0.08	30	40	8	50	0.15	
Ewes	0.51	0.24	0.65	0.15	0.2	0.14	25	30	9	40	0.15	
MTL[2]							300	1000	25	500	10	1000

[1]Contains food waste, soybean hulls and rolled corn in a 40:55:5 ration, wet weight basis.
[2]MTL = Maximum Tolerable Limit (NRC, 1980).

dietary requirements for pregnant sows, cows, and ewes, and to the NRC's MTL for livestock. These data suggest that the product extruded can meet these animals' requirements for potassium, magnesium, sulfur, zinc and possibly calcium without exceeding the MTL for any of the elements tested. These data suggest that food waste (even when extruded with soybean hulls and corn) comes closer to meeting the dietary requirements for ruminants than nonruminants. Calcium and phosphorus are more likely to be fulfilled for cows and ewes with this particular extruded feedstuff amended with food waste than for sows. However, with appropriate mineral supplementation the mixed product that was extruded could potentially satisfy the dietary requirements for protein, minerals, and energy for gestating sows. The ADF value of food waste and extruded food-waste amended feed appeared too high for food waste to serve as a primary feedstuff for growing-finishing swine if a high rate of performance (high average daily gain) is desired.

The selected fractionate composition of the extruded feed fed to lambs during a digestion trial is shown in table 11.19. The ether extract and acid detergent lignin values are similar to the values reported in table 11.17. The crude protein, acid detergent fiber, and cellulose fractionates are numerically higher than those reported in table 11.17. The values shown in table 11.17 represent all of the food-waste amended feed that was extruded, while the values in table 11.19 reflect only the extruded feed utilized in the digestion trial. Not all of the pulped food waste collected on a given day was extruded. A few containers of food waste were randomly selected from the total production for extruding because physical constraints limited extruded feed production to about 182 to 227 kg of total feed per day. Accordingly, any conclusions referencing the digestibility coefficients obtained should be relative to the true representiveness of the pulped food waste utilized to produce the extruded feed.

Table 11.20 reports the apparent digestibility coefficients of the fractionate selected for analysis. The protein and ether coefficients are higher than the coefficients for acid detergent fiber and cellulose. The variation in digestion coefficients between these fractionates is typical and the relative values obtained are reflective of the variation expected.

The digestible energy value obtained for the extruded feed (3.22 ± 1.50 Mcal:kg) can be compared to the reported digestible energy of other feedstuffs (NRC 1985) typically included in the diets of sheep. Alfalfa hay (IFN 1-00-054) has reported values ranging from 2.38 to 2.56 Mcal:kg depending on the stage of maturity when harvested. The digestible energy of the extruded feed in this study has a considerably higher value than that of alfalfa hay. Corn grain (IFN 4-02-931) has a

Table 11.19. Mean (±SD) percent selected fractionate composition of extruded feed fed during the digestion trial (dry matter basis)

Item	Dry matter	Protein	Ether extract	Acid detergent fiber	Cellulose	Ash	Acid detergent lignin
Mean	89.778	22.95	4.29	34.14	29.72	0.07	4.35
SD	17.280	8.64	0.85	6.48	6.93	0.24	0.54
n	28.000	28.00	28.00	28.00	28.00	28.00	28.00

Table 11.20. Mean (±SD) apparent digestibility coefficients of selected fractionates (%) and apparent digestible energy (Mcal:kg) of extruded feed[1] (dry matter basis)

Item	Dry matter	Protein	Ether extract	Acid detergent fiber	Cellulose	Apparent digestible energy
Mean	54.50	60.30	76.23	42.60	49.00	3.22
SD	2.68	15.68	3.77	4.39	3.57	1.50
n	19.00	19.00	19.00	19.00	19.00	19.00

[1]Extruded feed contains pulped food waste: soybean hulls: rolled shelled corn in a 40:55:5 ratio, wet weight basis.

reported digestible energy value of 3.84 Mcal:kg. Oat grain (IFN 4-03-309) has digestible energy values ranging from 2.91 - 3.40 Mcal:kg depending on its test weight classification in kg:hl (pounds:bushel). Brewers dehydrated grains (5-02-141) have a digestible energy value of 3.09 Mcal:kg. The digestible energy value of 3.22 ± 1.50 Mcal:kg for the extruded feed of this study is much higher than the value expected for forages but lower than that of corn grain. The digestibility coefficient of the extruded feed amended with pulped food waste is within that range cited for oat grain. However, the coefficient of variation (0.47) for the extruded feeds digestibility coefficient is high and suggests additional digestion trials should be conducted to more accurately assess the true digestibility of this extruded feed containing food waste.

Table 11.21 shows the body weight changes of the lambs consuming the extruded feed during the digestibility trial. The lambs had an acceptable average daily gain (0.33 kg) during the adaptation period. An average daily gain of this magnitude suggests that the extruded feed containing food waste is acceptable to the lambs. The lambs did lose a negligible amount of weight while in the metabolism cages during the collection period. A small weight loss during collection is not atypical.

Table 11.21. Mean (±SD) lamb body weight (kg) and change in weight at the beginning and end of the adaptation period and at the end of the collection period

| | Adaptation Period | | | Collection Period | |
	Initial Weight	Ending Weight	Weight Change	Ending Weight	Weight Change
Mean	52.30	58.96	6.66	57.87	−1.09
SD	7.26	7.00	1.20	4.60	0.84
n	21.00	21.00	21.00	21.00	21.00

Conclusions and Implications

The garbage recycling initiative at ISU is unique. Utilizing waste pulpers to process institutional pre- and postconsumer food waste places garbage in a physical form well suited for extruding. The pulping process reduces the moisture content of wet garbage creating a pulped food waste more suitable for extruding. However, the high moisture content of pulped food waste is still somewhat problematic but not unsolvable for extrusion technology to overcome. Mixing pulped food waste with dry feedstuffs such as soybean hulls and rolled shelled corn facilitate extruding. In general, the ruminant digestive system is well suited to utilize unprocessed vegetable food residuals and appropriately heat-treated food waste containing animal protein.

References

Federal Register, 1997.21 CFR part 589: Substances prohibited from use in animal food or feed; animal proteins prohibited in ruminant feed; final rule vol. 62, no. 108. US Department of Health and Human Services, Food and Drug Administration, Rockville, MD.

Flores, R. A., D. A. Ferris, M. K. King, and C. W. S. Shanklin. 1993. Characterization of food waste streams: a proximate analysis of plate and production wastes from university and military dining centers. Amer. Sci. Agri. Eng. Ann. Meetings. Chicago, IL.

Jurgens, M. H. 1993. Animal Feeding and Nutrition, 7th ed. P.93. Dubuque, IA: Kendall/Hunt Publishing Co.

Kelley, T. R., O. Pancorbo, W. Merka, S. Thompson, M. Cabrera, and H. Barnhart. 1994. Fate of selected bacterial pathogens and indicators in fractionated poultry litter during storage. *J. of Applied Poultry Research* 3: 279.

Kelley, T. R., O. Pancorbo, W. Merka, S. Thompson, M. Cabrera, and H. Barnhart. 1995. Bacterial pathogens and indicators in poultry litter during re-utilization. *J. of Applied Poultry Research* 4: 366.

Kornegay, E. T., G. Vander Noot, W. MacGrath, J. Welch, and E. Purkhiser. 1965. Nutritive value as a feed for swine. I. Chemical composition, digestibility, and nitrogen utilization of various types of garbage. *J. An. Sci.* 24:219.

Kornegay, E. T., G. Vander Noot, W. MacGrath, and K. Barth. 1968. Nutritive value as a feed for swine. III. Vitamin composition, digestibility, and nitrogen utilization of various types. *J. An. Sci.* 27: 1345.

Kornegay, E. T., G. W. Vander Noot, K. M. Barth, G. Garber, W. S. MacGrath, R. L. Gilbreath, and F. J. Bielk. 1970. Nutritive evaluation of garbage as a feed for swine. Bull. No. 829. College of Agriculture and Environmental Science. New Jersey Agricultural Experiment Station. Rutgers—the State University. New Brunswick, NJ.

Kris-Etherton, P. M., D. Kummel, M. E. Russel, D. Deron, S. Mackey, J. Borchers, and P. D. Wood. 1988. The effect of diet on plasma lipids, lipoproteins and coronary heart disease. *J. Am. Diet. Assoc.* 88:1373.

Kroyer, G. T. 1995. Impact of food processing on the environment—an overview. *Food Science and Technology* 28(6): 547.

Michigan Dairy Nutrition Manual. 1990.

Myer, R. O., T. A. DeBusk, J. H. Brendemuhl, and M. E. Rivas. 1994. Initial assessment of dehydrated edible restaurant waste (DERW) as a potential feedstuff for swine. Res. Rep. A1-1994-2. College of Agriculture. Florida Agricultural Experiment Station. University of Florida. Gainesville, FL.

NRC (National Research Council). 1980. Mineral Tolerance of Domestic Animals. Washington: National Academy Press.

NRC (National Research Council). 1984. Nutrient requirements for beef cattle. 6th Ed. Washington: National Academy Press.

NRC (National Research Council). 1985. Nutrient requirements for sheep. 6th Ed. Washington: National Academy Press.

Rust, S. 1991. Nutrient composition of byproduct and non-traditional feedstuffs. 52nd Minnesota Nutrition Conference.

Troeger, K., G. Reuter, and D. Schneider. 1983. Use of dried chicken manure in cattle feeding, from the aspects of meat hygiene and feed technology. I. Microorganisms, nutrients and residues in battery-hen manure and maize silage used as basal feed. *Berliner und Munchener Tierarzliche Wochenshrift* 96(11): 388-397.

U.S. Environmental Protection Agency (USEPA). 1997. Characterization of Municipal Waste in the US: 1996 Update, EPA/530-R-92-019. U.S. Printing Office, Washington, DC.

Walker, P. M. and A. E. Wertz. 1994. Analysis of selected fractionates of a pulped food waste and dish water slurry combination collected from university cafeterias. *J. Anim. Sci.* 72(Suppl.1):523(Abstr.).

Walker, P. and T. Kelley. 1997. Selected fractionated composition and microbiological analysis of institutional food waste, pre- and post-extrusion. Final Report, Illinois Council on Food and Agricultural Research (CFAR).

Walker, P. M., S. A. Wertz, and T. J. Marten. 1997. Selected fractionate composition and digestibility of an extruded diet containing food waste fed to sheep. Abst. *J. Anim. Sci.* 75(Suppl.1):253(Abstr.).

Wiltbank, J. N., W. W. Rowden, J. E. Ingalls, and D. R. Zimmerman. 1964. Influence of post-partum energy level on reproductive performance of Hereford cows restricted in energy intake prior to calving. *J. Anim. Sci.* 23:1049.

12

Concerns When Feeding Food Waste to Livestock

by Daniel G. McChesney

Regulations

The feeding of food waste to animals is subject to both federal and state regulations. At the federal level, the FDA has primary responsibility, but the USDA's Animal and Plant Health Inspection Service (APHIS) has responsibility for preventing the spread of disease among animals. Thus, the Swine Health Protection Act (1998) and the Poultry Improvement Plan (1998) are USDA-APHIS responsibilities. The states administer a variety of feed laws that in some cases are more stringent than federal requirements, one example being the feeding of garbage. As a general rule, state requirements, which are as stringent or more stringent than federal regulations, are permitted and generally will take precedent over federal law within the state. State feed control officials through the Association of American Feed Control Officials (AAFCO) have developed model feed regulations and numerous feed ingredient definitions, to which most states adhere (1999).

On October 4, 1997, all provisions of the BSE (bovine spongiform encephalopathy) regulation became effective (Animal Proteins Prohibited in Ruminant Feed 1998). This regulation, which is discussed in more detail later, prohibits the feeding of certain food waste to ruminants. Most food waste is still permitted to be fed provided certain cooking and recordkeeping requirements are met. AAFCO has recently given tentative status to definitions for restaurant food waste and food processing waste. These definitions appear in the Official Publication of AAFCO (1999) along with the existing definitions for Dehydrated Garbage and Dehydrated Food Waste and better reflect the actual products used.

As recycling becomes more of an environmental and economic necessity, the use of nontraditional sources of feed ingredients becomes

227

more appealing. The focus of this chapter will be on the use of food waste as a feed ingredient. For our purpose, food waste will include plate waste, food-processing waste, out-of-date and out-of-specification food, and by-products of the food-processing and food-serving industries. Within this broad grouping, we will concentrate on those items that have not been traditionally used or considered as feed ingredients and in general fit the AAFCO definitions for "Restaurant Food Waste"[1] and "Food Processing Waste."[2] Thus products such as dried bakery products, dried dairy products, and the protein and fat products produced by rendering will not be discussed.

This chapter is divided into three sections—the first looks at general safety concerns and evaluation criteria, the second looks at specific products, and the final section looks at other products, flocculating agents, and processes on which the FDA has been asked to comment. The last section is included to make the users of nontraditional feed ingredients aware of other products that could be in a feed product and of processes that might be used to produce the product. This last section also serves to point out that many items, other than what we commonly consider food, have nutritive value and thus the potential to be recycled and utilized in animal feeds.

Before beginning the discussion, it is important to emphasize that those involved in the processing are producing a food product and have an obligation to produce one that's safe and wholesome, and those using the product to feed animals must use the product in accordance with existing federal, state, and local regulations. Therefore, industry should review all the steps involved in producing the feed by-products and, most importantly, avoid the use of ingredients or processes that might result in poisonous or deleterious substances entering animal feed.

Concerns and Evaluation Criteria

Our principle concerns over the use of food waste, garbage, by-products, or coproducts are related to the poorly defined and variable nature of the products, the numerous sources, the collection vehicles/receptacles that can vary from dedicated trucks and receptacles to collection by the same trucks used to collect other waste products, the potential variety of contaminants, and the level of control (or lack of control) encompassed by local and state laws/regulations.

In addition to the concerns stated above, the processing of these potentially recyclable materials, can vary from minimal with no testing for contaminants to commercial processing with pesticide and chemical screens. Clearly, with all these variables, regulating this portion of the recycling industry is challenging. Regulating it other than on a case-by-case basis at the federal level is currently not feasible.

Case-by-case evaluations are currently made by the FDA regarding the safety of these products and their compliance with the Federal Food, Drug, and Cosmetic Act (FFDCA or the Act, 1998a). The safety evaluations are currently made using all available information. This information includes, but is not limited to, the source (i.e., restaurants, food processors, etc.); the contaminants present or likely to be present; the process used in processing the material for animal feed; screening procedures for detecting chemical, microbiological, and filth contaminants; collection methods/vehicles; and agreements between the supplier and processor outlining the supplier's responsibilities for preventing contamination of the recyclable material. The evaluations do not address whether the collection, disposal, and processing of these products comply with state, local, or other federal laws that might apply to waste disposal.

Contaminant Screens

As the FDA and the states gain information about specific products or classes of products, we are able to provide general guidance with regard to potential hazards associated with the products. For products falling into the broad category of food waste, a heating step of at least 180° F for 20 minutes would be adequate to address our microbiological concerns, and a method for detecting and removing metal, glass, etc. would largely address filth. Pesticide screens should, at a minimum, be able to detect halogenated (lindane, heptachlor, dieldrin, aldrin, etc.) and organophosphate (malathion, etc.) pesticides. Chemical screens should detect halogenated compounds, particularly polychlorinated biphenyls (PCBs). The particular pesticide and chemical screens will depend on the nature of the product. A commercial laboratory should be able to provide advice on the appropriate screens and methodology based on the starting material. We realize that human food products are not normally expected to be contaminated; however, since not all of the food products entering the food-waste chain may have been offered for human food, it is worth considering why they are being diverted to animal feed. In at least some cases, this will be because a tolerance or specification for the human food product was not met.

Food-Waste Products

The term "food waste" can encompass many products and have many different meanings. Dried bakery products, dehydrated garbage, dehydrated food waste, dried potato products, and pasta products are all recognized animal feed ingredients with AAFCO (1999)definitions and there are many more examples. Plate waste has been used by the FDA in the BSE regulation (Animal Proteins Prohibited in Ruminant Feed 1998)to describe "inspected meat products which have been cooked and offered for human food and further heat processed for feed." Plate waste also has been used by industry and academia to describe food products discarded by patrons of restaurants, cafeterias, fast-food establishments, theme parks, and institutions such as hospitals and prisons. The USDA uses the word "garbage" to describe what industry and academia refer to as plate waste. In responding to a USDA comment to the BSE regulation, the FDA was careful to differentiate between the FDA's (62 FR 30936, June 5, 1997) definition of plate waste and the USDA's (Swine Health Protection, Definitions in Alphabetical Order 1998)definition of garbage. For the purpose of the BSE regulation, plate waste and garbage are not the same thing. From the above, it is obvious that when discussing food waste, there are ample opportunities for misunderstanding.

For the purpose of our discussion and also from a practical approach, there are two classes of food-waste products, those with an AAFCO definition and those without a definition.

An AAFCO definition identifies products with regard to content, and in some instances, moisture. The identity allows the product to be used by a nutritionist in formulating a feed, indicates that the product has a history of safe use as a feed ingredient, and allows the marketing of the product while minimizing the potential for economic fraud. The AAFCO definitions applying to our broad category of food waste are listed in the AAFCO Official Publication (1999). Most definitions applying to food-waste products can be found in the miscellaneous products section or the milk products section.

Two current definitions, Dehydrated Food Waste[3] and Dehydrated Garbage[4], are very broad and seem to most closely fit industry's and academia's definition of plate waste. However, both definitions are consistent with the USDA definition of garbage, and products described by either of these definitions may or may not be exempt from the FDA's BSE regulation(1998). In addition to the legal considerations, the potential variability of the product makes it difficult to use in formulating a feed, difficult to market based on a protein, fat, and fiber specification, and increases the potential for safety problems. While these are challenges to using the product, they are not insurmountable. Normal business records showing from whom the product is obtained, contracts/agreements between the generator and processor indicating what material is acceptable for inclusion, documentation of training of personnel collecting the product, the labeling of holding/collection containers, and results of periodic monitoring could all be used to define and characterize this product.

In general, the FDA would not object to feeding a product meeting either of these definitions, provided it was not adulterated with substances that would result in animal health problems or produce unsafe residues in meat, milk, or eggs (i.e., pesticides, industrial chemicals, pathogenic microorganisms, drug residues, etc.) and it was used in accordance with all applicable regulations.

In 1999, the AAFCO definitions for Restaurant Food Waste[5] and Food-Processing Waste were given tentative status and placed in the AAFCO Official Publication. These new definitions more accurately describe the material actually comprising plate waste and food waste than the existing definitions for Dehydrated Food Waste and Dehydrated Garbage.

The following are examples of products that the FDA has commented on but do not have AAFCO definitions. These products are being used in animal feeds but on a limited basis while an AAFCO definition is being sought.

Specific Products

Grease

The FDA has not opposed the use in animal feed of fryer grease, restaurant grease, sludge, or products defined in the fats and oils section of the AAFCO Official Publication (1999) when it consists entirely of edible by-products used in, or obtained from, the preparation of human food.

However, the FDA is opposed to the use of sewer grease or any product that has come in contact with or passed through the same drain as sanitary sewer water or solid matter as a component of human or animal food. Furthermore, they also are opposed to the use of grease of unknown origin as a component of human or animal food. This position was affirmed through regulatory action in August 1990 in which a warning letter was issued to a Colorado firm and again in 1994 with warning letters to firms in Georgia and Alabama. The basis for issuance of the warning letters was that sewer grease and grease of unknown origin are unfit for food and therefore adulterated under the Act (FFDCA 1998b) because the potential contaminants could not be known with any certainty. The FDA position specifically addresses sewer grease and grease of unknown origin and should not be interpreted as unilaterally applying to all grease trap waste.

The use of grease trap waste from floor drains, pot wash drains, dishwasher drains, sink drains, etc. is opposed in animal feed *unless* the contaminants were known (or not present) and did not result in unsafe tissue residues in milk, meat, and eggs or present a health hazard to animals.

In addition to FDA guidance, the 1999 Official Publication of AAFCO includes a note in the fats and oils section and definitions addressing this issue.

Filter Cake and Biomasss

There is no objection by the FDA of filter cake material, generated from the manufacturing of cheeses and salad dressings, to be used in animal rations for swine, beef cattle, or poultry provided it contained only food grade chemicals and/or chemicals approved for use on food contact surfaces and it complies with the AAFCO (1999) definition of Dairy Food By-Products.

Biomass resulting from enzyme production is another potential feed source. The FDA is currently handling these situations on a case-by-case basis. In general, the Center has not objected to feeding nongenetically engineered biomass, provided food grade chemicals were used in the production process, and the organism used to produce the enzyme appears on the AAFCO (1999) direct fed microorganisms (DFM) list. With regard to inactivation of the organisms, it is not necessary to inactivate organisms that are on the DFM list prior to feeding. However, inactivation becomes an issue if the organism is genetically modified or has no history of food use. Biomass can have high levels of nitrogen, however, and upwards of one-third may be in the form of nucleic acids that may not be available to the animal. Limited experience with these products suggests ruminants handle the product better than nonruminants. Feeding high levels of nucleic acid may be associated with some toxicological problems in certain species.

Olestra

Olestra is an approved food additive for savory snacks provided appropriate amounts of vitamins A, D, E, and K are added to the snack. Olestra appears to prevent the "normal" absorption of fat-soluble vitamins. Olestra has been extensively studied in humans, rodents, and swine and appears to be almost completely indigestible. However, olestra is reported to cause ill-defined gastrointestinal upset in some humans at approved levels.

AAFCO (1999) has established a definition for products resulting from the acid hydrolysis of sucrose polyesters, (Hydrolyzed Sucrose Polyesters, Feed Grade; definition 33.15) such as olestra, to make them digestible. No AAFCO definition exists for the use of indigestible sucrose polyesters in feed, and there's no documentation to show that the recyclable olestra-based products (ROBP) are processed in a way that makes the olestra digestible (i.e., the vast majority of the sucrose-fatty acids bonds are cleaved).

The FDA has four concerns for the use of olestra-containing products. First, there is the potential need for supplementation with vitamins A, D, E, and K. Second, is the need for assurance that olestra is safe when fed to dogs, cats, ruminants, horses, and poultry. Also, there is the need for consistent labeling. Finally, a need may exist for the producers to know the decrease in caloric density that can be expected in olestra-containing products.

Recently, information was received on the effect of the inclusion of olestra-containing products in poultry and swine rations. The information was reviewed, and it was determined that provided certain limitations are followed, olestra-containing products can be safely included in the rations of swine and poultry. We are requesting information on the maximum percent of olestra and maximum percent of indigestible fat from olestra be provided on the label. This requirement could be met in various ways. One possibility is to guarantee the minimum amount of digestible fat in the product. The label of an ingredient that included olestra-containing products would need to have feeding instructions such that a complete ration containing the ingredient will contain no more than 1,000 ppm (0.1%) of olestra on a dry matter basis. The use of olestra-containing products could be expanded to other species as information becomes available.

Nontraditional Ingredients, Flocculating Agents, and Processes

Nontraditional Ingredients

The following products have been used in animal feed rations. While the products are not specifically within the scope of the food-waste definition, they serve to illustrate the many other products that have potential for inclusion in animal feeds. Distillers dried grains and manure/litter are covered by a variety of AAFCO definitions (1999) depending on the actual content of the product. Newsprint does not have an AAFCO definition and its use is mostly as a bedding material.

Distillers Dried Grains. In April 1994, at least 4,632 cattle from 7 feed-lots in Kansas became ill, and 706 died. Initial investigation showed that all the feedlots recently had purchased and fed milo distillers dried grains and solubles (MDDGS) from a local ethanol production facility. An on-farm epidemiological investigation by state and federal agencies showed that in addition to the MDDGS, all affected feedlots also were also feeding a particular ionophore antibiotic.

The investigation, coupled with analytical testing by the FDA's National Forensic Chemistry Center in Cincinnati, showed that the MDDGS were contaminated with several analogues of two macrolide antibiotics. A detailed inspection of the local ethanol production facility was conducted. The investigation showed that this facility had processed and distilled waste ethanol that contained two macrolide antibiotics and several analogues. This distillation took place immediately prior to the outbreak of the cattle deaths. After distillation, the solids from the waste ethanol were added to the MDDGS and caused the MDDGS to be con-taminated with several analogues of two macrolide antibiotics.

The USDA tested the meat from exposed animals in the feedlots, and the FDA tested the milk from dairies exposed to the contaminated MDDGS. No residues were detected in either case.

A feeding trial performed by Kansas State University (KSU) showed conclusively what the FDA suspected following the on-farm epidemio-logical investigation. In the feeding trial, the macrolide antibiotic con-taminated MDDGS in conjunction with FDA-approved amounts of the ionophore antibiotic reproduced the clinical signs and lesions noted in the field. The FDA-approved amounts of the ionophore antibiotic with uncontaminated MDDGS produced no ill effects and the macrolide antibiotic contaminated MDDGS without the ionophore antibiotic also produced no ill effects.

This example clearly illustrates the concerns of using waste products, by-products, or coproducts and the difficulty there will be in developing universal guidance to cover the numerous products likely to come to market in the next several years.

Manure. Animal waste products must be disposed of in a manner that does not endanger public health and is environmentally sound. There are currently four possible disposal methods—burying in landfills, incin-eration, land application, and feeding to animals after processing.

Burying in landfills is not a viable solution because of the volume of material involved, the limited amount of landfill space, and potential for run-off. Incineration is not currently viable, again because of the volume and limited incinerator capacity. The use of waste products as fuel is showing promise, but feasibility and technology are still developing.

Land application of waste as a fertilizer has and continues to be a reasonable option. However, recent problems with *Pfiesteria* in fishing areas of Maryland, Virginia, and North Carolina are causing the land application to be reassessed in these areas. Using the product as an animal feed ingredient after processing remains a viable option and one with which health or safety problems have not been associated when the product is adequately processed.

Recycled animal waste products, including chicken manure, contain significant nutritive value. These products can be fed provided they are correctly processed and free of unsafe microbial, chemical, and heavy metal contaminants. AAFCO (1999) has established safety criteria and definitions that these products must meet. Almost all the states follow this guidance. The FDA considers products not meeting these criteria and definitions to be adulterated and would so state this if a state were to take regulatory action or if the product was found in interstate commerce.

In summary, the disposal of animal waste products will continue to be a problem, to which there is not currently a single solution. A combination of feeding, land application, and incineration may offer the best solution for the immediate future.

Newsprint. The recycling of paper has led to questions concerning whether recycled newspaper, recycled paper, or paper by-products could be safely used as animal bedding or as a source of cellulose.

The FDA has objected to the use of newsprint as a feed ingredient and as a bedding material. Since bedding often is consumed by the animal, the basis for an objection is the same for both uses. The use of newsprint was initially objected to because of concerns that the process used in bleaching wood pulp could lead to dioxin residues in the paper, and that the ink may contain chemicals that could present an animal-safety problem or a residue problem in products destined for human consumption.

Over the last several years, the process of bleaching wood pulp has changed so that our concerns for dioxin contamination are minimal. However, the problem with inks still exists. Several newspapers have switched to vegetable-based inks and use FD&C approved dyes for color. It is recognized that there may be a number of cases in which the particular facts and controls may enable newsprint to be safely used. There continues to be a case-by-case review and opinion regarding the suitability of newsprint.

In addition to the safety concerns associated with the use of newsprint as a feed ingredient, an AAFCO definition for the product would need to be established prior to its use in commercial feeds.

Flocculating Agents

One source of food waste is food-processing plants. The by-products from food manufacturing can be directed to the animal feed market directly from the processing line or the fat and protein can be removed from the processing water prior to the water entering the municipal system or a stream. Several products can be used to remove and concentrate the fat and protein prior to it being used for feed. Two products that are used for this purpose are polyacrylamide and chitosan.

Polyacrylamide. Polyacrylamide is used as a flocculating agent for food-processing water streams. The flocculated product can become incorporated into animal feed through inclusion in the rendered product or by direct incorporation. The Center for Veterinary Medicine is currently reviewing the regulatory status of polyacrylamide when used as a flocculating agent in the waste water streams of food-processing plants and the subsequent use of the flocculate material as an animal feed. One company, a manufacturer of polyacrylamide, has submitted a GRAS petition for this use. The GRAS petition has not yet been filed. CVM has not concurred in the company's GRAS finding and is currently resolving the scientific and legal issues raised by this use of polyacrylamide. Manufacturers and users of polyacrylamide should be aware that CVM could conclude that when used in this fashion, polyacrylamide is a food additive that requires a food additive regulation before it can be legally used. As part of its review, CVM is exploring whether to propose such a regulation. Until the status of this use of polyacrylamide is resolved, CVM does not anticipate pursuing enforcement actions.

Chitosan. Chitosan is another product that, as a 2% chitosan solution, is used as a waste water coagulant for products intended for animal feed. The FDA does not object to the use of a chitosan solution for this purpose provided the chitosan level does not exceed 4 ppm chitosan. At this usage level and coupled with the fact that flocculated material from the waste water stream of food-processing plants is not fed as the only component of a complete animal feed ration, we were able to conclude that this use of chitosan was acceptable under the AAFCO definition (1999) for chitosan.

Processes

Food waste can be processed by several methods involving heat or a combination of heat and pressure. In many cases, the species for which the ingredient is intended will determine the method that can be used. Methods that are commonly used are rendering, cooking in either open or closed containers, extrusion, pelleting, and composting. An important consideration in choosing a method is whether specific regulations apply because of the species fed the product.

Two regulations that impact the decision on processing are the Swine Health Protection Act (1998) and the BSE regulation (Animal Proteins Prohibited in Ruminant Feed 1998).

The Swine Health Protection Act requires that garbage (food waste for our purpose) intended to be fed to swine shall be heated throughout at boiling (212° F or 100° C at sea level) for 30 minutes. Garbage shall be agitated during cooking, except in steam cooking equipment, to ensure that the prescribed cooking temperature is maintained throughout the cooking container for the prescribed length of time. This requirement does not apply if the garbage is to be fed to other animal species.

The BSE regulation does not specifically address the feeding of garbage, rather it uses plate waste as an example of "inspected meat products which have been cooked and offered for human food and further heat processed for feed." The distinction between garbage and plate waste is that plate waste does not contain any uncooked meat products, whereas garbage might. A product meeting the further heat-processing requirements of the BSE regulation can be fed to ruminants.

The BSE regulation does not specifically address what constitutes heat processing. There are a variety of commercial processes and various temperatures that could be used to meet the heat-processing requirement. The following discussion looks at the five processes mentioned above with respect to the Swine Health Protection Act (1998) and the BSE regulation (Animal Proteins Prohibited for Use in Ruminant Feed 1998).

Rendering, which uses either the batch process or the continuous process, meets the requirements of both the Swine Health Protection Act and the BSE regulation. Rendering temperatures normally are in the range of 240 to 300° F, with times ranging from 30 minutes to several hours. A thorough discussion of rendering systems used in the United States can be found in the preamble to the proposed BSE regulation (62 Federal Register 564-565, January 3, 1997).

Cooking in an open or closed container, provided the internal temperature of the product reaches 212° F and is maintained for 30 minutes, the requirements of both the Swine Health Protection Act and the BSE regulation are met.

Dry extrusion at 284° F for approximately 30 seconds in a system that has pressure differential of approximately 40 atm meets the requirements of the BSE regulation. Other extrusion processes that are operated at temperatures above 200° F for extended dwell times also meet the BSE regulation requirements. Extrusion processes that operate at 212° F with dwell times of 30 minutes could meet the Swine Health Protection Act requirements. Extrusion processes that fail to have the product reach an internal temperature of at least 212° F for 30 minutes do not meet the Swine Health Protection Act.

Pelleting does not meet the requirements of the Swine Health Protection Act. Pelleting processes, in which the internal temperature of the product in the conditioner exceeds 200° F and the dwell time is such that the total heat energy is similar to that of the extrusion process or the heating requirements of the Swine Health Protection Act, can meet the further heat processing requirements of the BSE regulation.

Composting of food waste (plate waste, garbage) does not meet the temperature requirements of either the Swine Health Protection Act or the BSE regulation. Composting of food waste that does not contain meat products from mammals would be an acceptable method of processing the food waste for feeding to ruminants or other animals other than swine. Examples of products that are not subject to the BSE regulation and thus could be composted are all vegetables, milk (cheese), poultry, fish, and bakery products. Products that are a mixture of various products but contain some mammalian meat products are subject to the BSE regulation and thus could not be composted.

The above discussion addresses the processing of food waste from the federal level. As mentioned in the chapter on regulating food waste, federal requirements generally do not supersede state or local requirements if these requirements are more stringent than the federal requirements.

References

Animal Proteins Prohibited in Ruminant Feed. 1998. Title 21 Code of Federal Regulations §589.2000. U.S. Government Printing Office. 1998.

Association of American Feed Control Officials. (AAFCO). 1999. Official Publication.

Federal Food, Drug, and Cosmetic Act as Amended. (FFDCA). 1998a. Department of Health and Human Services, Food and Drug Adminstration.

Federal Food, Drug, and Cosmetic Act as Amended. (FFDCA). 1998b. § 402 (a)(3). Department of Health and Human Services, Food and Drug Administration.

62 Federal Register 564-565, January 3, 1997.

62 Federal Register 30936, June 5, 1997.

Poultry Improvement Plan. 1998. Title 9 Code of Federal Regulations §. 145 and 147. U.S. Government Printing Office.

Swine Health Protection, Definitions in Alphabetical Order. 1998. Title 9 Code
of Federal Regulations §166.1. U.S. Government Printing Office.

Swine Health Protection Act. 1998. Title 9 Code of Federal Regulations § 166.
U.S. Government Printing Office.

NOTES

1. **Restaurant Food Waste** is composed of edible food waste collected
from restaurants, cafeteria, and other institutes of food preparation.
Processing and / or handling must remove any and all undesirable constituents including crockery, glass, metal, string and similar materials.
The guaranteed analysis shall include the maximum moisture, unless
the product is dried by artificial means to less than 12% moisture and
designated as "Dehydrated Restaurant Food Waste." If part of the grease
and fat is removed it must be designated as "Degreased."

2. **Food Processing Waste** is composed of any and all animal and vegetable products from basic food processing. This may include manufacturing or processing waste, cannery residue, production over-run, and
otherwise unsaleable material. The guaranteed analysis shall include the
maximum moisture, unless the product is dried by artificial means to
less than 12% moisture and designated as "Dehydrated Food Processing
Waste." If part of the grease and fat is removed it must be designated as
"Degreased."

3. **Dehydrated Food Waste.** Any and all animal and vegetable produce picked up from basic food processing sources or institutions where
food is processed. The produce shall be picked up daily or sufficiently
often so that no decomposition is evident. Any and all undesirable constituents shall be separated from the material. It shall be dehydrated to
a moisture content of not more than 12% and be in a state free from all
harmful microorganisms.

4. **Dehydrated Garbage** is composed of artificially dried animal and
vegetable waste collected sufficiently often that harmful decomposition
has not set in, and from which have been separated crockery, glass,
metal, string and similar materials. It must be processed at a temperature sufficient to destroy all organisms capable of producing animal
diseases. If part of the grease and fat is removed, it must be designated
as "Degreased Dehydrated Garbage."

5. **Restaurant Food Waste** is composed of edible food waste collected
from restaurants, cafeteria, and other institutes of food preparation.
Processing and/or handling must remove any and all undesirable constituents including crockery, glass, metal, string and similar materials
and provide product in a state free of all harmful microorganisms. The

guaranteed analysis shall include the maximum moisture, unless the product is dried by artificial means to less than 12% moisture and designated as "Dehydrated Restaurant Food Waste." If part of the grease and fat is removed it must be designated as "Degreased."

13

Rendering Food Waste

by Don A. Franco and Gary Pearl

Introduction

The concept and fundamentals of recycling have emerged as a logical option in the United States in the last 25 years. During the past 10 years, it has accelerated into an extensive, nationwide undertaking — a type of mantra and commitment to transform and process/reprocess all applicable waste products into usable materials, thus circumventing the need of further burdening our extended municipal waste disposal system (Burnham 1996). This heightened trend is somewhat ironic but readily understood because historically different forms of recycling have been practiced by mankind for more than 2,000 years. The obvious questions are, Why the current increased use of a known and accepted practice? and, What are the future opportunities, including economic benefits for expanded options of recycling, particularly into new areas of commercial applicability? An educated answer could be improvised from the country's food dynamics.

The United States has the most abundant and diversified food supply in the world, where consumers can choose from an average of 50,000 different food products on a typical visit to the supermarket (Kantor et al. 1997). The large quantities of unconsumed edible food and other losses in every facet of the food production and marketing systems, including plate waste from institutions and food-service establishments, underscore the possibilities for further utilization of these types of products. The losses include all the meats, bread, and other foods prepared by restaurants and food distributors that are never served, including blemished and overripe products, which for cosmetic reasons may not be marketable. This fact, plus the new and changing environmental laws and regulations throughout the country force considerate evaluations of options other than traditional waste disposal methods existing. This will provide opportunities to entrepreneurs in the United States to evaluate the potential concurrent financial rewards for investing in a novel

241

business venture like rendering and recycling of food waste into usable finished products.

About 20% of the food produced for human consumption in the country is wasted — enough to supply nearly one-fifth of the population with their daily caloric needs. Preliminary research suggests that food losses exceeded 100 billion pounds in 1995 (Kantor et al. 1997), implying substantial economic and environmental costs to society in the form of land, energy, labor, water, and other resources used to produce food not eaten. The concurrent costs for waste disposal and environmental degradation resulting from landfills and incinerators aggravate the complex and diversified issues associated with food-waste processing and recycling. The problem is compounded by the fact that estimates of the amount of food waste in the United States lack reliable and comprehensive data for major commodities, thus precluding an accurate assessment of potentially recoverable raw material that can be utilized.

The Economic Importance of Feed for Livestock and Poultry

Feed for beef cattle constitutes the greatest single cost of their production, accounting for 65 to 75% of the total cost of maintaining beef cows. In turn, this is a powerful influence on cow fertility and calf weaning weight — the two biggest factors determining success in the cattle business. Feed is also a major item of expense in finishing cattle, accounting for 70 to 80% of the cost of feedlot finishing, exclusive of the purchase price of the animals (Ensminger et al. 1990a). In dairy cattle, feed determines the productivity and profitability and accounts for about 55% (range 45 to 65%) of the cost of milk production. Since the price of milk and the cost of feed move independently of each other, good managers are continually challenged to reduce cost and still optimize their feeding programs (Ensminger et al. 1990b). Similarly, feed accounts for 65 to 75% of the total cost of producing pork, and swine producers endeavor to provide rations that are nutritionally satisfactory and inexpensive — in essence, least-cost rations that optimize production of quality pork per unit of feed consumed. Swine production research shows that nutritional deficiencies contribute to reproductive failures and baby pig mortality. Of all commercial farm animals, pigs are most likely to suffer from mineral deficiencies because the skeleton of the pig supports greater weight in proportion to its size than that of any other farm animal (Ensminger et al. 1990c).

The efficient use of feed is of paramount importance to poultry producers. Feed varies between 55 and 75% of total production cost of poultry. Poultry feeding has changed more than the feeding of any other species. In 1940, it required 4.7 lbs of feed to produce 1 lb weight gain in broilers; in 1998, it required only 1.8 lbs of feed to produce 1 lb of weight gain (Ensminger et al. 1990d). The dramatic changes in the poultry industry over the past 25 years have led to remarkable efficiency of production and a relatively stable market. Consumption of poultry has increased approximately 150% over the past 30 years in part due to development of numerous innovative products. Obviously, feed cost is an integral and significant factor in production and profitability of meat. This fact provides opportunities for entrepreneurs to assess new or changing options to supply a safe feed with high nutritive value and is affordable to every sector of animal agriculture, including niche markets such as aquaculture and companion animal food. The latter markets are now multibillion dollar-enterprises with opportunities for continued expansion and growth.

The Role of Biosecurity in Processing Food Residuals

Feeding trials indicate dehydrated edible restaurant waste has potential as a feed source for swine. Concurrently, official publications and bulletins from USDA-APHIS affirm that feeding food waste has the potential to help producers reduce costs, but there are inherent risks associated with the practice (USDA-APHIS 1995). Although swine was referenced as an exemplar, a degree of relative risk exists whenever livestock are fed human food residuals, unless appropriate precautions are incorporated. The government's risk analysis for swine highlighted the microbial risk factors and examined both the relevance for disease transmission to livestock and an associated link to public health. The pathogens highlighted were *Toxoplasma*, *Trichinella*, *Salmonella*, and *Campylobacter*. The bulletins were published prior to the existing food safety initiatives promulgated by the government and the spiraling concerns of consumer advocates who are critical of government food safety policies. Equally important is the significance of the farm environment, feed, and husbandry to the total food cycle — that is, the "farm to fork" analogy that mandates the principles of biosecurity in the processing of food residuals.

The Center for Veterinary Medicine (CVM) of the FDA regulates the level of contaminants permitted in animal feed to ensure that the food for animals and humans is safe. Adulteration is basically defined as (1) a feed (food) that bears or contains any poisonous or deleterious substance that may render it injurious to health; (2) if it is otherwise unfit for food (feed); (3) if it bears or contains any food additive that is unsafe (not approved) within the meaning of Section 409 of the Federal Food, Drug, and Cosmetic Act as amended (see FFDCA 1998). Fundamentally, the industry has the responsibility for producing an unadulterated product. The role of government is oversight to assure that the industry is responsible and accountable in meeting its obligations to produce a safe food (feed).

An initial problem with the entire subject involves negative terminology with terms like waste and garbage in lieu of more considerate options. The public has and will likely continue to have serious reservations about feeding food waste. The obvious inference is that if food waste is garbage, it ought to be treated as such. This has a degree of legitimacy, but in this era of producers examining the use of alternative and accessible nontraditional raw material to reduce feed cost and enhance economic benefits, it is in our best interests that the entire subject be evaluated objectively.

Challenges and Considerations

The potential economic, environmental, and resource conservation attributes and the variety of possible food residuals that could be used for animal feed are growing rapidly, providing incentives for processing into usable finished products. Nonetheless, an array of challenges and concerns should be addressed including

(1) The nature and variety of the available products, including the wide ranges of sources, demand a standard for handling and processing. Thus, there should be different requirements for dried bakery waste in contrast to those for a variety of meats.

(2) Contaminants and infectious agents will vary according to the origin and nature of the raw material.

(3) The varying levels of control or current lack of control encompassed by different jurisdictions (city, county, state, federal) could be inconsistent, overlapping, and problematic.

(4) The possible transfer of residues or pathogens to livestock tissue that could ultimately affect animal or human health must be considered a risk.

(5) A standard operating procedure must be established to detect contaminants such as heavy metals, pesticides, or pathogens.

(6) A standard operating procedure must be established for time-temperature processing, depending on raw material to assure finished product safety.

(7) A standard operating procedure must be established to separate food residuals from packaging and other nonedible materials such as plastic, aluminum, and glass.

(8) A memorandum of agreement should assure cooperation and understanding between the generator of the food residuals, the hauler, the processor, and the farmer.

(9) The product must be consistent from batch to batch so that users may predict its nutritive value and performance.

(10) Nutrient content must be established for each batch.

(11) Anticipating variable nutritive value, develop guidelines for inclusion rates for specific rations for the major livestock species.

(12) Establish who is liable for contaminations.

(13) Create collaborations for research and feeding trials, nutritional analyses, safety and environmental factors to maintain credibility for alternative feeds.

(14) License processing facilities to ensure minimum acceptable standards for sanitation and hygiene in processing of food residuals to livestock feed.

(15) Industry must provide the leadership and advocacy for food residual processing and feeding, including the promulgation of rules.

At the outset, we should decide whether the hazard analysis and critical control points (HACCP) program is applicable to the recycling and processing of food residuals. HACCP may provide the means to systematically examine each step in the recycling process as to ensure food safety from farm to table.

Processing

Several companies manufacture equipment with claims to process food-waste streams so as to yield pathogen-free, sterilized, shelf-stable feed ingredients and animal feeds through a system of air drying with uniform heating, drying, and cooling of the product. Although most claim sterility of the end product, these claims should be validated scientifically. Given the diversity of food residuals, at the very least we must assure microbial safety of alternative feeds if they are to be seriously considered. Guaranteed time-temperature processing assurances to accomplish this objective are a mandate. This is a challenge to everyone interested in the complexities associated with processing nontraditional feeds for commercial acceptance.

Another major concern expressed by many potential processors is the moisture (water) content of many of the available residuals. High water content will adversely affect the resulting product yield. This obviously impacts economic considerations and determines whether the investment will provide meaningful returns.

In summary, food waste can be categorized as originating from four main sources: raw material wastes, food-processing wastes, postprocessing wastes, and postconsumer wastes (Lencki 1995). The following six major considerations are prerequisites for food residual recycling or feeding:

(1) A steady supply/availability of raw material

(2) Microbial, physical, and chemical safety

(3) Cost effectiveness as a feed ingredient including handling, processing, transportation, and nutritive value

(4) Consistency of raw material from batch to batch

(5) Quality as a feed material, protein and fat values, and effect on meat quality

(6) Minimal environmental impacts from containers, wrapping, and secondary disposal concerns including sanitation and hygiene.

Discussion

Livestock and poultry rations are formulated to meet the specific nutrient requirements of each species consistently from day to day. The potential use of alternative feeds like food waste (residuals), either as a feed ingredient or the major component of the ration, requires addressing the concerns of both efficacy and safety. Doubtless, the proper use of food residuals can offer a good value to livestock and poultry feeders and other likely users such as the aquaculture and companion animal food industries. To gain mobility in the marketplace, these products must be cost competitive on a nutrient content basis, have reasonable storage times or shelf life, be free from contaminants and/or adulterants, and should not require supplementation with other feed ingredients. These alternative products also must be produced in a "transparent" environment using the fundamental principles of biosecurity to assure product safety.

Summary

Animal feeding and nutrition constitutes the largest single cost of livestock production by far. This provides opportunities for utilization of nontraditional sources of feed that could minimize cost and provide nutritive value that is safe to livestock and acceptable to producers. The emerging concerns of food safety and the philosophical shift for establishing controls from "farm to fork" to assure a safe food supply create challenges for every sector of production. They especially are relevant for food residuals' use because of the association with waste or garbage, and the perceived concomitant problems inherent to the product(s). Thus, we all have very distinct responsibilities to form alliances to encourage research, determine nutritive values, assure process safety, and form an alliance of subject matter experts who can deal with the complexities associated with all phases of production. The many unknowns and variables will continue to stretch our ingenuity and mandate collaboration to synthesize answers.

References

Burnham, F. 1996. The Rendering Industry: An Historical Perspective. *In:* The Original Recyclers, Franco, D.A. and Swanson, W. (eds.), APPI/FPRF/NRA Publishers. Pages 1-15.

Ensminger, M. E., J. E. Oldfield, and W. W. Heinemann. 1990a. (eds.) Feeding Beef Cattle. *In:* Feeds and Nutrition. Clovis, CA: Ensminger Publishing Company. Pages 690-806.

Ensminger, M. E., J. E. Oldfield, and W. W. Heinemann. 1990b. (eds.) Feeding Dairy Cattle. *In:* Feeds and Nutrition. Clovis, CA: Ensminger Publishing Company. Pages 807-872.

Ensminger, M. E., J. E. Oldfield, and W. W. Heinemann. 1990c. (eds.) Feeding Swine. *In:* Feeds and Nutrition. Clovis, CA: Ensminger Publishing Company. Pages 951-1008.

Ensminger, M. E., J. E. Oldfield, and W. W. Heinemann. 1990d. (eds.) Feeding Poultry. *In:* Feeds and Nutrition. Clovis, CA: Ensminger Publishing Company. Pages 1009-1064.

Federal Food, Drug, and Cosmetic Act as Amended. 1998. Department of Health and Human Services, Food and Drug Administration.

Kantor, L. S., K. Lipton, A. Manchester, and V. Oliveira. 1997. Estimating and Addressing America's Food Losses. Economic Research Service, U.S. Department of Agriculture Report. *Food Review.* 20: 3-11.

Lencki, R. W. 1995. Issues and Solutions for Recycling Food Wastes. Symposium—Recycled Feeds for Livestock and Poultry. March 1995. Ontario, Canada.

United States Department of Agriculture, APHIS, Veterinary Services: Risk of feeding food waste to swine: Public health diseases. Centers for Epidemiology and Animal Health Bulletin. April 1995. Fort Collins, CO.

14

Concerns With the Use of Nontraditional Feed Wastes and By-products

by Perry J. Durham

Introduction

Most people forget that items we use have to go somewhere when you're done with them; nothing disappears. Given the status of landfills and the costs of incineration, the agriculture and food-service communities would be well-advised to create a useful fate for their residuals.

Many people think of range animals when they think of livestock — cattle on the plains with cowboys caring for them. In my veterinary practice in the Northeast, we dealt with mostly small family farms with a lot of dairy, a few pigs, a few chickens, and a few horses. I suspect that most Americans, the typical American consumer, still have this impression. The problem with that image is that it doesn't really fit with the modern agriculture developing in this country.

Contextual Reference

Farm Business, Not Farming

We are witnessing the rise of a whole new class of agriculture — integrated production agribusiness. These new business entities aren't driven by lifestyle choices or by the need to live on a farm to care for animals. They are driven by profitability. For instance, Premium Standard Farms, Murphy Family Farms, or Tyson, Inc. all are driven by bottom lines and profitability issues. They must manage product flow to optimize profitability. They manage production flow of animals. Rather than widgets from a factory, essentially they are managing a type of protein production. They aim toward a consistent product to satisfy consumers who equate inconsistency with poor quality. That in turn drives a new paradigm into the feed industry. The mentality is no longer simply getting a good buy on some material and using it somewhere on the farm. Agribusinesses require a consistently high quality input supply stream at the lowest available cost.

Consumerism

At the same time, the modern retail consumer has arisen as another class that we don't understand well. We are all a part of this consumer group, but we sometimes miss this point simply because we are so close to it. These days one often finds homes where both Mom and Dad are working or at least running, and the kids often are running somewhere in the middle. Increasingly, this is a typical, modern American family.

This family has some huge concerns about whatever they purchase and increasingly huge concerns about the safety of products they purchase. Is it OK to run to the store and pick up a bag of horse feed? Will it be safe, not only for the horse, but also for the kids who might check it out?

Zero Risk

In the best of all worlds, we would enjoy no risk of anything bad happening. Maybe this sounds idealistic, but that's what most consumers expect — a zero-risk scenario. Did you know that half the cost of a push-type lawnmower is to cover liability? For example, someone might lift it and use the lawnmower as a hedge trimmer. The manufacturer lost in such a court case because specific written instructions to not use it to trim your hedges were not provided.

That kind of world can scare people more than product regulatory issues. I can participate in developing regulatory controls, and I feel good about that because the regulations have inherent value. For example, the FDA's good manufacturing practices are also sound business practices. However, the arena of product liability is frightening as some very large sums of money have been awarded to plaintiffs.

Public Education

In terms of a safety issue, what is it in the consumer's mind? What he knows about safety is what he's heard or read in the press, seen from an excerpt of a scientific paper in a magazine, heard on the radio, or heard from friends and neighbors.

Now compare this learning method with what is known about consumer responses. According to various consumer research groups, if a person encounters a product that makes them feel satisfied or content, they will typically tell 7 to 11 people that they received good value from it. If they weren't satisfied with the product or experience, they will tell 19 to 29 people that they were not satisfied. This gives some insight into the consumer mindset. Generally, consumers don't talk about food safety in a positive light. If they address the subject of safety, generally it is with a negative tone.

Legalities

Every business must deal with federal, state, county, and often local municipal regulations. In addition, most levels of government are represented by multiple agencies (e.g., FDA, Environmental Protection Agency, and Occupational Safety and Hazard Administration from the federal government and departments of agriculture, natural resources, environments, etc. from the various states). Add to these concerns about nuisance and tort civil litigation, and most businesses try to avoid anything new.

Regulatory

Under the broad federal regulatory apparatus, as the public becomes increasingly concerned about food safety, the issue of animal feed safety arises. For example, public concerns over Creutzfeld-Jakob disease and bovine spongiform encephalopathy have led to the restriction of feeding ruminant-derived protein products to ruminants. The primary charge for regulations is consumer safety, not a bad goal for industry as well as government. Industry needs satisfied consumers.

New regulations usually result from consumer complaints translated by legislators through the regulatory agencies into regulatory proposals. This process also provides industry, as well as individuals, opportunities to participate in the creation and evolution of regulations at all levels. Can we work with consumers to create regulations that would protect consumers and at the same time serve industry needs?

Liability

Doing a good job complying with the regulations is not good enough. We are a litigious society. If you market a product, you must address the inherent liability concerns. If the customer is not happy for whatever reason, you may face uncomfortable litigation. This costs businesses large and small many dollars, not to mention the physical and psychological stress on employees. The concept of risk drives this issue. We in the feed manufacturing industry must address this issue. How the ultimate retail consumer views your product has a big impact on how our customers view our product, and how or whether we can utilize food-waste products.

An Example — Farmland Industries Feed Division

The following is a review of how Farmland Industries Feed Division addresses the issue of ingredient inputs utilized in manufacturing animal feeds.

The Process and Flow Chart

Product Acquisition. Typically at Farmland, the nutritionists are expected to create and investigate new ingredients. They may have seen literature about a new feed ingredient, or they may have had discussions with peers. Occasionally, Farmland production people initiate an ingredient change after they learn of an improved production process such as improved pelletability or flowability. On other occasions, Farmland field staff may discover a local purchasing opportunity. However, this issue drives less and less of feed manufacturing. Instead, what drives our ingredient decisions is that our customers demand better quality and higher consistency. Possibly the biggest concern about using a new ingredient is whether or not we have some solid data on its performance characteristics.

Product Specifications. We prefer data that has been published in refereed scientific journals. This gives us insight as to how the ingredient will perform and hopefully contains typical analytical ranges of nutrient composition. Farmland is fortunate to have its own research and development farm and a feed testing laboratory, so the characteristics of new ingredients can be tested at Farmland when published data are not available. This is normally an extensive process before adopting a new ingredient, unless there is a considerable amount of documentation available.

Governmental Regulations. The Federal Food, Drug and Cosmetics Act states there should not be any poisonous or deleterious substances in feed. Is there really any way that you can absolutely guarantee there is nothing poisonous to any given species a person chooses to expose feed material to? Take starch for example. Most people would consider carbohydrates as safe feed. However, if you suddenly expose a ruminant to an unlimited supply of starch, you can kill that animal pretty easily. Is that a poisonous substance? How do you define poisonous? Most feed ingredients would be considered Generally Recognized As Safe (GRAS) under federal regulation.

Logistics and Transportation. Several years ago, the Brazilian government mandated that their domestic citrus industry stop dumping their waste products. Because the citrus processors had to do something with the

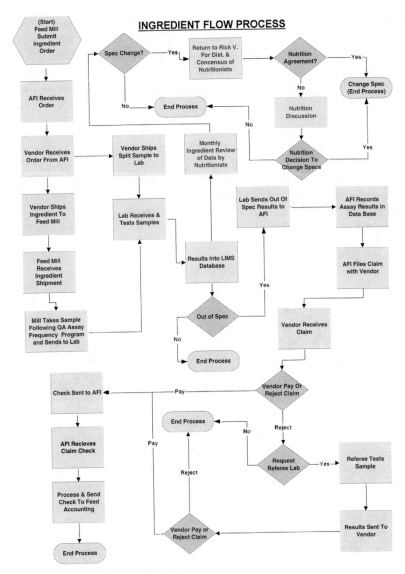

Figure 14.1. Farmland Industries feed ingredient flow process.

waste, low cost dried citrus pulp is now available as a feed ingredient in Florida and Texas. This is a good example of a product that we know well; we know how it performs in animal models, and we know the costs of alternatives. That is, what other ingredients can be substituted for citrus pulp, and at what price level will they move in or out of a given formula?

Least-cost Formulation. The Farmland Feed Division has a group of about eight people whose main role is to reformulate feed rations. One person's role is to play "what if." If I pull this ingredient out and put another in, what does this do to our cost of manufacturing? Will it "price in" at certain mills and not "price in" at other mills? What levels of the ingredient will be needed at those mills? Does this level justify its purchase and transport? All of these activities go on essentially every day.

The Farmland Ingredient Process

The flow chart illustrates our "ingredient process." A Farmland buying group purchases ingredients such as meat and bone meal, citrus pulp, corn, etc. from vendors around the country. Sampling the ingredient starts as soon as the shipments are delivered. Some samples are sent to the testing labs directly from the vendors, depending upon whether or not there are trading rules established for the particular commodity. This gets back to the tort issues versus the regulatory issues. Samples also come from the production facilities to the feed lab.

The lab analyzes the samples for the purchase specifications (usually protein, fiber, calcium, etc.) The analytical data are filed into the Farmland database, the Laboratory Information Management System. Analytical results are compared to specifications to determine compliance. When the results are within the purchase specifications, the data are filed for use by the nutritionists to adjust the formulation database. When the product does not meet the purchase specifications, the out-of-specification information is forwarded to a claims manager who files a claim against the vendor who supplied the product.

Farmland can not tolerate the potential losses from ingredient variability. Profit margins are generally so slim that all the value must be captured in the ingredients purchased. Some organizations have gone to online, real-time generation of this information. Basically they have a near infrared analysis unit at their production facility, to perform analyses on ingredients as they enter their facility for unloading. Those data are then fed directly into their computer, so that all their diets are reformulated every night based on the latest data.

Farmland is trying to capture all the value in each purchase of every ingredient. Ingredient variability destroys the process and robs the company of profit potential. If there is excessive variability in one ingredient nutrient, some other nutrient(s) must be over-formulated to compensate for the variability of the first one. That's why we demand of our suppliers that their product be consistent.

This is a very collaborative effort. Nutritionists, veterinarians, formulation specialists, mill personnel, and purchasing people all are involved in determining whether an ingredient is actually useful, where it might be useful, and the kind of levels that would be useful.

Summary

Least-cost formulation is the bottom line, but there is also the need to satisfy the customer. Most feed businesses know what type of animal performance they can expect if they supply the proper inputs. Cost consciousness is huge. However, there are areas where people will not tolerate safety or performance compromises, and these have become prominent issues.

The public is becoming increasingly aware of bacterial contamination of food and feed, such as *Salmonella*, *E. coli*, and *Campylobacter*. These hot issues drive regulatory changes by the FDA and USDA. We, as consumers, want everything totally safe. Issues like microbial resistance are addressed almost daily on the local news programs or national talk shows. Will these concerns from the consumer permit us to make use of food waste?

An augury is a fortuneteller. They pick up the gut, run it through their hands, and read the liver. Perhaps that may predict the regulatory future for the feed industry as well as any other measure. We live in tumultuous, changing times, but this gives us an opportunity to affect the outcome.

Appendixes

The Association of American Feed Control Officials, Incorporated
Sharon Senesac, Assistant Secretary/Treasurer
P.O. Box 478
Oxford, IN 47971
Web Site: *http://www.aafco.org/*

U.S. Food and Drug Administration
Office of Surveillance and Compliance
Center for Veterinary Medicine, HFV-220
Division of Animal Feeds
7500 Standish Place
Rockville, MD 20855
Web Site: *http://www.fda.gov/cvm/*

U.S. Department of Agriculture
Animal and Plant Health Inspection Services (APHIS)
Veterinary Services
Center for Animal Health
4700 River Road, Unit 43
Riverdale, MD 20737
Web Site: *http://www.aphis.usda.gov/*

Canada
Canadian Food Inspection Agency
Animal Health and Production Division
Feed Section
59 Camelot Drive
Nepean, Ontario, Canada K1A 0Y9
Web Site: *http://www.cfia-acia.agr.ca/*

U.S. State Feed Control Officials

Alabama	Agricultural Commodities Inspection Division
	Department of Agriculture and Industries
	Richard Beard Building
	P.O. Box 3336
	Montgomery, AL 36109-0336
Alaska	Alaska Department of Natural Resources
	Division of Agriculture
	1800 Glenn Highway, Suite 12
	Palmer, AK 99645
Arizona	Arizona Department of Agriculture
	Environmental Services Division
	1688 West Adams
	Phoenix, AZ 85007
Arkansas	Division of Feed and Fertilizer
	State Plant Board
	1 Natural Resources Drive
	Little Rock, AR 72205
California	California Department of Food and Agriculture
	Agriculture Commodities and Regulatory Services Branch
	1220 N Street-Room A 472
	Sacramento, CA 95814-5621
Colorado	Colorado Department of Agriculture
	Feed Control
	2331 W. 31st Avenue
	Denver, CO 80211
Connecticut	Connecticut Department of Agriculture
	State Office Building, Rm. 291
	165 Capitol Avenue
	Hartford, CT 06106

Delaware	Delaware Department of Agriculture Division of Consumer Protection 2320 South DuPont Highway Dover, DE 19901
Florida	Florida Department of Agriculture and Consumer Services, Feed Section 3125 Conner Boulevard, ME-2 Tallahassee, FL 32399-1650
Georgia	Georgia Department of Agriculture Plant Food, Feed, and Grain Division Capitol Square Atlanta, GA 30334
Hawaii	Hawaii Department of Agriculture Commodities Branch P.O. Box 22159 Honolulu, HI 96823-2159
Idaho	Idaho Department of Agriculture Bureau of Feeds and Plant Services P.O. Box 790 Boise, ID 83701
Illinois	Department of Agriculture Bureau of Agricultural Products Inspection Fairgrounds, P.O. Box 19281 Springfield, IL 62794
Indiana	Office of the Indiana State Chemist Purdue University Feed Administrator 1154 Biochemistry Building West Lafayette, IN 47907-1154
Iowa	Iowa Department of Agriculture Feed Bureau Wallace State Office Building Des Moines, IA 50319
Kansas	Kansas Department of Agriculture Division of Inspections 901 S. Kansas Avenue B 7th Floor Topeka, KS 66612-1272
Kentucky	Division of Regulatory Services Room 103, Regulatory Services Building University of Kentucky, Lexington, KY 40546-0275

Louisiana	Louisiana Department of Agriculture and Forestry Division of Agricultural Chemistry P.O. Box 25060, University Station Baton Rouge, LA 70894-5060
Maine	Department of Agriculture, Food, and Rural Resources Division of Quality Assurance and Regulations State House Station No. 28 Augusta, ME 04333
Maryland	Maryland Department of Agriculture 50 Harry S. Truman Parkway Annapolis, MD 21401
Massachusetts	Massachusetts Department of Food and Agriculture Bureau of Farm Products Leverett Saltonstall Building 100 Cambridge St. Boston, MA 02202
Michigan	Michigan Department of Agriculture Feed and Drug Coordinator P.O. Box 30017 Lansing, MI 48909
Minnesota	Minnesota Department of Agriculture Agronomy and Plant Protection Division 90 West Plato Boulevard St. Paul, MN 55107
Mississippi	Mississippi Department of Agriculture Feed, Fertilizer, and Lime Division P.O. Box 1609 Jackson, MS 39215-1609
Missouri	Missouri Department of Agriculture Plant Industries Division Bureau of Feed and Seed P.O. Box 630 Jefferson City, MO 65102-0630
Montana	Montana Department of Agriculture Agricultural Sciences Division P.O. Box 200201 Helena, MT 59620-0201

Nebraska	Nebraska Department of Agriculture Bureau of Plant Industry P.O. Box 94756 Lincoln, NE 68509
Nevada	Nevada Division of Agriculture Bureau of Plant Industry 350 Capitol Hill Avenue Reno, NV 89502
New Hampshire	New Hampshire Department of Agriculture, Markets, and Food 25 Capitol Street P.O. Box 2042 Concord, NH 03302-2042
New Jersey	New Jersey Department of Agriculture Bureau of Agricultural Chemistry Division of Regulatory Services CN30 Trenton, NJ 08625
New Mexico	Agricultural and Environmental Services Bureau of Feed, Seed, and Fertilizer P.O. Box 30005, Dept. 3150 Las Cruces, NM 88003-0005
New York	Department of Agriculture and Markets Division of Food Safety and Inspection Capitol Plaza-I Winners Circle Albany, NY 12235
North Carolina	North Carolina Department of Agriculture Food and Drug Protection Division 4000 Reedy Creek Road Raleigh, NC 27606
North Dakota	North Dakota Department of Agriculture Registration 600 E. Blvd., 6th Floor Bismarck, ND 58505-0020
Ohio	Ohio Department of Agriculture Reynoldsburg Laboratory Divisions Feed and Fertilizer Section Reynoldsburg, OH 43068-3399
Oklahoma	Oklahoma Department of Agriculture Plant Industry and Consumer Services 2800 N. Lincoln Boulevard Oklahoma City, OK 73105-4298

Oregon

Oregon Department of Agriculture
Feed Specialist
635 Capitol Street NE
Salem, OR 97310-0110

Pennsylvania

Pennsylvania Department of Agriculture
Bureau of Plant Industry
Division of Agronomic Services
2301 N. Cameron Street
Harrisburg, PA 17110-9408

Puerto Rico

Puerto Rico Department of Agriculture
Apartado 10163
Santurce, PR 00908-1163

Rhode Island

Rhode Island Division of Agriculture
Department of Environmental Management
235 Promenade Street
Providence, RI 02908-5767

South Carolina

South Carolina Department of Agriculture
Laboratory Division
Registration Officer
P.O. Box 11280
Columbia, SC 29211-1280

South Dakota

South Dakota Department of Agriculture
Division of Agricultural Services
Office of Agronomy Services
523 East Capitol B Foss Building
Pierre, SD 57501-3182

Tennessee

Tennessee Department of Agriculture
Division of Regulatory Services
P. O. Box 40627
Nashville, TN 37204

Texas

Office of the Texas State Chemist
Feed and Fertilizer Control Service
P.O. Box 3160
College Station, TX 77841-3160

U.S. Virgin Islands

U. S. V. I. Department of Agriculture
Estate Lower Love
Kingshill, VI 00850

Utah

Utah Department of Agriculture
350 N. Redwood Road
Box 146500
Salt Lake City, UT 84114-6500

Vermont	Plant Industry, Laboratory, and Standards Division 116 State Street, Drawer 20 Montpelier, VT 05620-2901
Virginia	Virginia Department of Agriculture and Consumer Resources Office of Product and Industry Standards P.O. Box 1163 Richmond, VA 23209
Washington	Washington Department of Agriculture Registration Services, Feed and Fertilizer Program P.O. Box 42589 Olympia, WA 98504-2589
West Virginia	West Virginia Department of Agriculture Regulatory Protection Division 1900 Kanawha Boulevard, E. Charleston, WV 25305
Wisconsin	Wisconsin Department of Agriculture Trade and Consumer Protection Agricultural Resource Management Division P.O. Box 8911 Madison, WI 53708
Wyoming	Wyoming Department of Agriculture Technical Services 2219 Carey Avenue Cheyenne, WY 82002-0100

Appendix B

The 1998 Amended Swine Health Protection Act

—CITE— 7 USC CHAPTER 69 — SWINE HEALTH PROTECTION

TITLE 7 — AGRICULTURE

Sec.
3801. Congressional findings and declaration of purpose.
3802. Definitions.
3803. Prohibition of certain garbage feeding; exemption.
3804. Permits to operate garbage treatment facility.
    (a) Application; issuance.
    (b) Cease and desist orders; suspension or revocation orders; judicial review.
    (c) Automatic revocation.
3805. Civil penalties.
    (a) Assessment by Secretary.
    (b) Judicial review.
    (c) Collection action by Attorney General.
    (d) Payment into United States Treasury.
    (e) Compromise, modification, or remittance.
3806. Criminal penalties.
3807. General enforcement provisions.
    (a) Injunctions.
    (b) Access to premises or facility and books and records; examination; samples.
    (c) Additional powers.
3808. Cooperation with States.
3809. Primary enforcement responsibility.
    (a) State obligation.
    (b) Inadequate enforcement or administration by State; termination of responsibility by Secretary.
    (c) Request of State official.
    (d) Emergency conditions.
3810. Repealed.
3811. Issuance of regulations; maintenance of records.
3812. Authority in addition to other laws; effect on State laws.
3813. Authorization of appropriations.

The Congress hereby finds and declares that—

(1) Raw garbage is one of the primary media through which numerous infectious or communicable diseases of swine are transmitted;

(2) If certain exotic animal diseases, such as foot-and-mouth disease, African swine fever, hog cholera, and swine vesicular diseases, gain entrance into the United States, such diseases may be spread through the medium of raw or improperly treated garbage which is fed to swine;

(3) African swine fever, which is potentially the most dangerous and destructive of all communicable swine diseases, has been confirmed in several countries of the Western Hemisphere, including the Dominican Republic, Haiti, and Cuba;

(4) Swine in the United States have no resistance to any of such exotic diseases and in the case of African swine fever there is a particular danger because there are no effective vaccines to this deadly disease;

(5) All articles and animals which are regulated under this chapter are either in interstate or foreign commerce or substantially affect such commerce, and regulation by the Secretary and cooperation by the States and other jurisdictions as contemplated by this chapter are necessary to prevent and eliminate burdens upon such commerce, to effectively regulate such commerce, and to protect the health and welfare of the people of the United States;

(6) The interstate and foreign commerce in swine and swine products and producers and consumers of pork products could be severely injured economically if any exotic animal diseases, particularly African swine fever, enter this country;

(7) It is impossible to assure that all garbage fed to swine is properly treated to kill disease organisms unless such treatment is closely regulated;

(8) Therefore, in order to protect the commerce of the United States and the health and welfare of the people of this country, it is necessary to regulate the treatment of garbage to be fed to swine and the feeding thereof in accordance with the provisions of this chapter.

—SOURCE—

(Pub. L. 96-468, Sec. 2, Oct. 17, 1980, 94 Stat. 2229.)

SHORT TITLE

Section 1 of Pub. L. 96-468 provided: "That this Act (enacting this chapter) may be cited as the 'Swine Health Protection Act'."

Sec. 3802. Definitions

For purposes of this chapter—

(1) The term "Secretary" means the Secretary of Agriculture;

(2) The term "garbage" means all waste material derived in whole or in part from the meat of any animal (including fish and poultry) or other animal material, and other refuse of any character whatsoever that has been associated with any such material, resulting from the handling, preparation, cooking, or consumption of food, except that such term shall not include waste from ordinary household operations which is fed directly to swine on the same premises where such household is located;

(3) The term "person" means any individual, corporation, company, association, firm, partnership, society, or joint stock company or other legal entity; and

(4) The term "State" means the fifty States, the District of Columbia, Guam, Puerto Rico, the Virgin Islands of the United States, American Samoa, the Commonwealth of the Northern Mariana Islands, and the territories and possessions of the United States.

—SOURCE—

(Pub. L. 96-468, Sec. 3, Oct 17, 1980, 94 Stat. 2229; Pub. L. 96-592, title V, Sec. 511, Dec. 24, 1980, 94 Stat. 3451.)

AMENDMENTS

1980 — Par. (4). Pub. L. 96-592 added par. (4). Sec. 3803. Prohibition of certain garbage feeding; exemption

(a) No person shall feed or permit the feeding of garbage to swine except in accordance with subsection (b) of this section.

(b) Garbage may be fed to swine only if treated to kill disease organisms, in accordance with regulations issued by the Secretary, at a facility holding a valid permit issued by the Secretary, or the chief agricultural or animal health official of the State where located if such State has entered into an agreement with the Secretary pursuant to section 3808 of this title or has primary enforcement responsibility pursuant to section 3809 of this title. No person shall operate a facility for the treatment of garbage knowing it is to be fed to swine unless such person holds a valid permit issued pursuant to this chapter. The Secretary may exempt any facility or premises from the requirements of this section whenever the Secretary determines that there would not be a risk to the swine industry in the United States.

—SOURCE—

(Pub. L. 96-468, Sec. 4, Oct. 17, 1980, 94 Stat. 2230.) Sec. 3804. Permits to operate garbage treatment facility

(a) Application; issuance any person desiring to obtain a permit to operate a facility to treat garbage that is to be fed to swine shall apply therefore to (1) the Secretary, or (2) the chief agricultural or animal health official of the State where the facility is located if such State has entered into an agreement with the Secretary pursuant to section 3808 of this title or has primary enforcement responsibility pursuant to section 3809 of this title, and provide such information as the Secretary shall by regulation prescribe.

No permit shall be issued unless the facility —

(1) Meets such requirements as the Secretary shall prescribe to prevent the introduction, or dissemination of any infectious or communicable disease of animals or poultry, and

(2) Is so constructed that swine are unable to have access to untreated garbage of such facility or material coming in contact with such untreated garbage.

(b) Cease and desist orders; suspension or revocation orders; judicial review whenever the Secretary finds, after notice and opportunity for a hearing on the record in accordance with sections 554 and 556 of title 5, that any person holding a permit to operate a facility to treat garbage in any State is violating or has violated this chapter or any regulation of the Secretary issued hereunder, the Secretary may issue an order requiring such person to cease and desist from continuing such violations or an order suspending or revoking such permit, or both. Any person aggrieved by an order of the Secretary issued pursuant to this subsection may, within sixty days after entry of such order, seek review of such order in the appropriate United States court of appeals in accordance with the provisions of sections 2341, 2343 through 2350 of title 28, and such court shall have jurisdiction to enjoin, set aside, suspend (in whole or in part), or to determine the validity of the Secretary's order. Judicial review of any such order shall be upon the record upon which the determination and order are based.

(c.) Automatic revocation the permit of any person to operate a facility to treat garbage in any State shall be automatically revoked, without action of the Secretary, upon the final effective date of the second conviction of such person pursuant to section 3806 of this title.

—SOURCE—

(Pub. L. 96-468, Sec. 5, Oct. 17, 1980, 94 Stat. 2230.)

SECTION REFERRED TO IN OTHER SECTIONS

This section is referred to in section 3806 of this title. Sec. 3805. Civil penalties

(a) Assessment by Secretary any person who the Secretary determines, after notice and opportunity for a hearing on the record in accordance with sections 554 and 556 of title 5, is violating or has violated any provision of this chapter or any regulation of the Secretary issued hereunder, other than a violation for which a criminal penalty has been imposed under this chapter, may be assessed a civil penalty by the Secretary of not more than $10,000 for each such violation. Each offense shall be a separate violation. The amount of such civil penalty shall be assessed by the Secretary by written order, taking into account the gravity of the violation, degree of culpability, and history of prior offenses; and may be reviewed only as provided in subsection (b) of this section.

(b) Judicial review the determination and order of the Secretary with respect thereto imposing a civil penalty under this section shall be final and conclusive unless the person against whom such an order is issued files application for judicial review within sixty days after entry of such order in the appropriate United States court of appeals in accordance with the provisions of sections 2341, 2343 through 2350 of title 28, and such court shall have jurisdiction to enjoin, set aside, suspend (in whole or in part), or to determine the validity of the Secretary's order. Judicial review of any such order shall be upon the record upon which the determination and order are based.

(c) Collection action by Attorney General if any person fails to pay a civil penalty under a final order of the Secretary, the Secretary shall refer the matter to the Attorney General, who shall institute a civil action to recover the amount assessed in any appropriate district court of the United States. In such collection action, the validity and appropriateness of the Secretary's order imposing the civil penalty shall not be subject to review.

(d) Payment into United States Treasury all penalties collected under authority of this section shall be paid into the Treasury of the United States.

(e) Compromise, modification, or remittance the Secretary may, in his discretion, compromise, modify, or remit, with or without conditions, any civil penalty assessed under this chapter.

—SOURCE—

(Pub. L. 96-468, Sec. 6, Oct. 17, 1980, 94 Stat. 2231.)

Sec. 3806. Criminal penalties

(a) Whoever willfully violates any provision of this chapter or the regulations of the Secretary issued hereunder shall be guilty of a misdemeanor and shall be fined not more than $10,000, or imprisoned not more than one year, or both.

(b) Any person who fails to obey any order of the Secretary issued under the provisions of section 3804 of this title, or such order as modified—

    (1) After the expiration of the time allowed for filing a petition in the court of appeals to review such order, if no such petition has been filed within such time; or

    (2) After the expiration of the time allowed for applying for a writ of certiorari, if such order, or such order as modified, has been sustained by the court of appeals and no such writ has been applied for within such time; or

    (3) After such order, or such order as modified, has been sustained by the courts as provided in section 3804(b) of this title; shall on conviction be fined not more than $10,000, or imprisoned for not more than one year, or both. Each day during which such failure continues shall be deemed a separate offense.

—SOURCE—

(Pub. L. 96-468, Sec. 7, Oct. 17, 1980, 94 Stat. 2231.)

SECTION REFERRED TO IN OTHER SECTIONS

This section is referred to in section 3804 of this title.

Sec. 3807. General enforcement provisions

(a) Injunctions the Attorney General, upon the request of the Secretary, shall bring an action to enjoin the violation of, or to compel compliance with, any provision of this chapter or any regulation issued by the Secretary hereunder by any person. Such action shall be brought in the appropriate United States district court for the judicial district in which such person resides or transacts business or in which the violation or omission has occurred or is about to occur. Process in such cases may be served in any judicial district wherein the defendant resides or transacts business or wherever the defendant may be found.

(b) Access to premises or facility and books and records; examination; samples any person subject to the provisions of this chapter shall, at all reasonable times, upon notice by a duly authorized representative of the Secretary, afford such representative access to his premises or facility and opportunity to examine the premises or facility, the garbage there at, and books and records thereof, to copy all such books and records and to take reasonable sample of such garbage.

(c) Additional powers for the efficient execution of the provisions of this chapter, and in order to provide information for the use of Congress, the provisions (including penalties) of sections 46 and 48 through 50 of title 15, are made applicable to the jurisdiction, powers, and duties of the Secretary in enforcing the provisions of this chapter and to any person subject to the provisions of this chapter, whether or not a corporation. The Secretary, in person or by such agents as he may designate, may prosecute any inquiry necessary to his duties under this chapter in any part of the United States.

—SOURCE—

(Pub. L. 96-468, Sec. 8, Oct. 17, 1980, 94 Stat. 2232.)

Sec. 3808. Cooperation with States

In order to avoid duplication of functions, facilities, and personnel, and to attain closer coordination and greater effectiveness and economy in administration of this chapter and State laws and regulations relating to the feeding of garbage to swine, the Secretary is authorized to enter into cooperative agreements with State departments of agriculture and other State agencies charged with the administration and enforcement of such State laws and regulations and to provide that any such State agency which has adequate facilities, personnel, and procedures, as determined by the Secretary, may assist the Secretary in the administration and enforcement of this chapter and regulations hereunder. The Secretary is further authorized to coordinate the administration of this chapter and regulations with such State laws and regulations whenever feasible: Provided, That nothing herein shall affect the jurisdiction of the Secretary under any other Federal law, or any authority to cooperate with State agencies or other agencies or persons under existing provisions of law, or affect any restrictions upon such cooperation.

—SOURCE—

(Pub. L. 96-468, Sec. 9, Oct. 17, 1980, 94 Stat. 2232.)

SECTION REFERRED TO IN OTHER SECTIONS

This section is referred to in sections 3803, 3804 of this title.

Sec. 3809. Primary enforcement responsibility

(a) State obligation

For purposes of this chapter, a State shall have the primary enforcement responsibility for violations of laws and regulations relating to the treatment of garbage to be fed to swine and the feeding thereof during any period for which the Secretary determines that such State —

(1) Has adopted adequate laws and regulations regulating the treatment of garbage to be fed to swine and the feeding thereof which laws and regulations meet the minimum standards of this chapter and the regulations hereunder: Provided, That the Secretary may not require a State to have laws that are more stringent than this chapter;

(2) Has adopted and is implementing adequate procedures for the effective enforcement of such State laws and regulations; and

(3) Will keep such records and make such reports showing compliance with paragraphs (1) and (2) of this subsection as the Secretary may require by regulation. Except as provided in subsection (c) of this section, the Secretary shall not enforce this chapter or the regulations hereunder in any State which has primary enforcement responsibility pursuant to this section.

(b) Inadequate enforcement or administration by State; termination of responsibility by Secretary whenever the Secretary determines that a State having primary enforcement responsibility pursuant to this section does not have adequate laws or regulations or is not effectively enforcing such laws or regulations, the Secretary shall notify the State. Such notice shall specify those aspects of the administration or enforcement of the State program that are determined to be inadequate. The State shall have ninety days after receipt of the notice to correct any deficiencies. If after that time the Secretary determines that the State program remains inadequate, the Secretary may terminate, in whole or in part, the State's primary enforcement responsibility under this chapter.

(c) Request of State official

(1) In general on request of the Governor or other appropriate official of a State, the Secretary may terminate, effective as soon as the Secretary determines is practicable, the primary enforcement responsibility of a State under subsection (a) of this section. In terminating the primary enforcement responsibility under this subsection, the Secretary shall work with the appropriate State official to determine the level of support to be provided to the Secretary by the State under this chapter.

(2) Reassumption nothing in this subsection shall prevent a State from reassuming primary enforcement responsibility if the Secretary determines that the State meets the requirements of subsection (a) of this section.

(d) Emergency conditions nothing in this section shall limit the authority of the Secretary to enforce this chapter whenever the Secretary determines that emergency conditions exist that require immediate action on the part of the Secretary and the State authority is unwilling or unable adequately to respond to the emergency.

—SOURCE—

(Pub. L. 96-468, Sec. 10, Oct. 17, 1980, 94 Stat. 2233; Pub. L.104-127, title IX, Sec. 914(a), Apr. 4, 1996, 110 Stat. 1186.)

AMENDMENTS

1996–Subsecs. (c), (d). Pub. L. 104-127 added subsec. (c) and redesignated former subsec. (c) as (d).

SECTION REFERRED TO IN OTHER SECTIONS

This section is referred to in sections 3803, 3804 of this title.

Sec. 3810. Repealed. Pub. L. 104-127, title IX, Sec. 914(b)(1),

Apr. 4, 1996, 110 Stat. 1186

Section, Pub. L. 96-468, Sec. 11, Oct. 17, 1980, 94 Stat. 2233, authorized Secretary to appoint and consult with advisory committees concerning matters within scope of this chapter.

Sec. 3811. Issuance of regulations; maintenance of records the Secretary is authorized to issue such regulations and to require the maintenance of such records as he deems necessary to carry out the provisions of this chapter.

—SOURCE—

(Pub. L. 96-468, Sec. 11, formerly Sec. 12, Oct. 17, 1980, 94 Stat.2233; renumbered Sec. 11, Pub. L. 104-127, title IX, Sec. 914(b)(2), Apr. 4, 1996, 110 Stat. 1186.)

PRIOR PROVISIONS

A prior section 11 of Pub. L. 96-468 was classified to section 3810 of this title prior to repeal by Pub. L. 104-127.

Sec. 3812. Authority in addition to other laws; effect on State laws the authority conferred by this chapter shall be in addition to authority conferred by other statutes. Nothing in this chapter shall be construed to repeal or supersede any State law prohibiting the feeding of garbage to swine or to prohibit any State from enforcing requirements relating to the treatment of garbage to be fed to swine or the feeding thereof which are more stringent than those under this chapter or the regulations hereunder.

—SOURCE—

(Pub. L. 96-468, Sec. 12, formerly Sec. 13, Oct. 17, 1980, 94 Stat. 2233; renumbered Sec. 12, Pub. L. 104-127, title IX, Sec. 914(b)(2), Apr. 4, 1996, 110 Stat. 1186.)

PRIOR PROVISIONS

A prior section 12 of Pub. L. 96-468 was renumbered section 11 and is classified to section 3811 of this title.

Sec. 3813. Authorization of appropriations there are hereby authorized to be appropriated such sums as may be necessary to carry out the provisions of this chapter.

—SOURCE—

(Pub. L. 96-468, Sec. 13, formerly Sec. 14, Oct. 17, 1980, 94 Stat.2234; renumbered Sec. 13, Pub. L. 104-127, title IX, Sec. 914(b)(2), Apr. 4, 1996, 110 Stat. 1186.)

PRIOR PROVISIONS

A prior section 13 of Pub. L. 96-468 was renumbered section 12 and is classified to section 3812 of this title.

NOTE: In 1996, Public Law 104-127 amended the Swine Health Protection Act. The conference report described the changes as follows:

2)   Swine health protection, Mount Pleasant National Scenic Area, and pseudorabies eradication program

The Senate amendment authorizes the Secretary, upon request of the Governor or other appropriate official of a State, to terminate the State's primary enforcement responsibility under the Swine Health Protection Act. This section also deletes the requirement that an advisory committee be appointed to evaluate state programs regulating the treatment of garbage to be fed to swine. (Section 544)

The House bill contains no comparable provision. The Conference substitute adopts the Senate provision with an amendment regarding the designation of the Mount Pleasant National Scenic Area and extending the Pseudorabies eradication program through 2002. (Section 914, 915, and 916)

TEXT OF AMENDMENT:

SEC. 914. SWINE HEALTH PROTECTION.

(a) Termination of State Primary Enforcement Responsibility.—Section 10 of the Swine Health Protection Act (7 U.S.C. 3809) is amended—
(1) by redesignating subsection (c) as subsection (d); and
(2) by inserting after subsection (b) the following:
(c) Request of State Official.—
(1) In general.—On request of the Governor or other appropriate official of a State, the Secretary may terminate, effective as soon as the Secretary determines is practicable, the primary enforcement respon-

sibility of a State under subsection (a). In terminating the primary enforcement responsibility under this subsection, the Secretary shall work with the appropriate State official to determine the level of support to be provided to the Secretary by the State under this Act.

(2) Reassumption.—Nothing in this subsection shall prevent a State from reassuming primary enforcement responsibility if the Secretary determines that the State meets the requirements of subsection (a).".(b) Advisory Committee.—The Swine Health Protection Act is amended—(1) by striking section 11 (7 U.S.C. 3810); and (2) by redesignating sections 12, 13, and 14 (7 U.S.C. 3811, 3812, and 3813) as sections 11, 12, and 13, respectively.

General Reference List for Research and Utilization

Allee, G. L. and R. H. Hines. 1972. Influence of fat level and calorie:protein ratio on performance of young pigs. *J. Anim. Sci.* 35:210. (Abstr.)

Altizio, B. A., P. A. Schoknecht, and M. L. Westendorf. 1998. Growing swine prefer a corn/soybean diet over dry, processed food waste. *J. Anim. Sci.* 76(Suppl.1):185. (Abstr.).

Ammerman, C. B. and P. R. Henry. 1991. Citrus and vegetable products for ruminant animals. Proc. Alternative Feeds for Dairy and Beef Cattle Sym. (Jordan, E. R., Ed.) Coop. Ext., Univ. Of Missouri, Columbia. pp. 103-110.

Animal Proteins Prohibited in Ruminant Feed. 1998. Title 21 Code of Federal Regulations '589.2000. U.S. Government Printing Office.

AOAC. 1990. Official Methods of Analysis (15th Ed.). Association of Official Analytical Chemists, Arlington, VA.

Association of American Feed Control Officials. 1999. *Official Publication.* Paul Bachman, ed. St. Paul, Minnesota.

Barth, K. M., G. W. Vander Noot, W. S. MacGrath, and D. R. Reynolds. 1966. Nutritive value of garbage as a feed for swine. II. Mineral content and supplementation. *J. Anim. Sci.* 25:52-57.

Bath, D., J. Dunbar, J. King, S. Berry, and S. Olbrich. 1998. Byproducts and unusual feedstuffs. *In:* Feedstuffs Ref. Issue 70(30):32-38. The Miller Publishing Co., Minnetonka, MN.

Belyea, R. 1991. Alternative Feeds: Chemical Composition. Proc. National Invitational Symposium for Alternatives Feeds for Dairy and Beef Cattle. E.R. Jordan, Ed. University of Missouri, Columbia. page 153.

Blake, J. P., M. E. Cook, and D. R. Reynolds. 1991. Extruding poultry farm mortalities. Am. Soc. Agric. Engr. Internat'l. meeting. Paper No. 914049. Am. Soc. Agric. Engr., St. Joseph, MI. USA.

Brandt, R. C. and K. S. Martin. 1994. The Food Processing Residual Management Manual. PA Dept. of Env. Resources Publ. No. 2500-BK-DER-1649.

Briggs, G., W. D. Gallup, V. G. Heller, A. E. Darlow and F.B. Cross. 1947. The Digestibility of Dried Sweet Potatoes by Steers and Lambs. Stillwater, OK, Oklahoma Agricultural Experiment Station, Oklahoma Agricultural and Mechanical College, Technical Bulletin No. T-28.

275

Burnham, F. 1996. The Rendering Industry: An Historical Perspective. *In:* The Original Recyclers, Franco, D.A. and W. Swanson. (eds.), APPI/FPRF/NRA Publishers. Pages 1-15.

Callis, J. J., P. D. McKetcher, and M. S. Shanahan. 1975. Foot-And-Mouth Disease, *In:* Disesases of Swine, 4th edition; Iowa State University Press.

CAST. 1995. Waste Management and Utilization in Food Production and Processing. Council for Agricultural Science and Technology. October, 1995. Ames, IA.

Childs, T. 1952. The history of foot-and-mouth disease in Canada. Proc. 56th Ann. Meeting. U.S. Livestock Sanit. Assoc., p. 153.

Clark, W. S. 1979. Our industry today: Whey processing and utilization, major whey product markets - 1976. *J. Dairy Sci.* 62:96-98.

Congressional Quarterly Weekly Report Vol. 38 (Oct.-Dec.1980), p.2889-3680.

Congressional Record, vol.126, p. 2733-4176.

Crickenberger, R. and R. E. Carawan. 1991. Using Food Processing By-products for Animal Feed. Raleigh, NC, North Carolina Cooperative Extension Service Fact Sheet, #CD-37.

Current Good Manufacturing Practice for Medicated Feeds. 1998. Title 21 Code of Federal Regulations 225. U.S. Government Printing Office.

Current Good Manufacturing Practice in Manufacturing, Packing, or Holding Human Food. 1998. Title 21 Code of Federal Regulations ' 110. U.S. Government Printing Office.

Derr, Donn A. "Economics of Food Waste Recycling", Presentation to the Conference of Recycling Economics and Marketing Strategies, December 1991.

Derr, D. A., A. T. Price, J. L. Suhr, and A. J. Higgins. 1988. Statewide system for recycling food waste. Biocycle *J. Waste Recyc.* 29(5):58-63.

Derr, D.A. 1991. Expanding New Jersey's Swine Industry Through Food Waste Recovery. Proceedings of Getting the Most From Our Materials Conference: Making New Jersey the State-of-the Art. L. Gilbert and B. Salas Ed. New Brunswick, NJ, USA. pp. 61-63, 76-77.

Dollar, K. K. 1998. The use of dried restaurant food residual products as a feedstuff for swine. M.S. Thesis. Univ. of Florida, Gainesville.

Doster, A., F.E. Mitchell, R.L. Farrell, and B.J. Wilson. 1978. Effects of 4-Ipomeanol, a product from mold-damaged sweet potatoes, on the bovine lung. *Vet Pathol* 15: 367-375.

Dukes, P. D., M. G. Hamilton, A. Jones, J. M. Schalk. 1987. Sumor, a multi-use sweetpotato. *HortScience* 22(1): 170-171.

Dunne, H. W., Leman, A.D. 1975. Diseases of Swine; Iowa State University Press, 4th edition.

Eastridge, M. L. (ed.). 1995. Proc. of Second National Alternative Feeds Symposium USDA and Univ. of Missouri-Columbia. Sept. 24-26, St. Louis, MO.

Eligibility for Classification as Generally Recognized as Safe (GRAS). 1998. Title 21 Code of Federal Regulations '570.30. U.S. Government Printing Office.

Engel, R. W., C. C. Brooks, C. Y. Kramer, D. F. Watson, and W. B. Bell. 1957. The composition and feeding value of cooked garbage for swine. Va. Agr. Exp. Sta. Tech. Bul. 133.

Ensminger, M. E., J. E. Oldfield, and W. W. Heinemann. 1990. Feeds and Nutrition. Second Revised Ed. The Ensminger Publishing Co. Clovis, CA.

Espinola C, N. 1992. Alimentacion animal con batata (Ipomoea batatas) en Latinoamerica. Turrialba 42(1): 114-126.

Federal Register. 50 Federal Register 27294-27296, July 2, 1985.

Federal Register. 62 Federal Register 30936, June 5, 1997.

Federal Register. 62 Federal Register 564-565, January 3, 1997.

Federal Register. 62 Federal Register 40570-40600, July 29, 1997.

Federal Food, Drug, and Cosmetic Act as Amended. 1998. ' 402 (a)(3). Department of Health and Human Services, Food and Drug Adminstration.

Federal Food, Drug, and Cosmetic Act as Amended. ' 201 (f). 1998. Department of Health and Human Services, Food and Drug Adminstration.

Federal Food, Drug, and Cosmetic Act, As Amended. United Sates Code, Title 21, et. al., U. S. Government Printing Office, Superintendent of Documents, Mail Stop: SSOP, Washington, DC 20402-9328.

Federal Meat Inspection Act (21 U.S.C. 601 et seq.).

Ferko, B.L, M.H. Poore, J.R. Schultheis, and G.M. Rogers. 1998. Feeding potato and sweetpotato by-products to beef cattle. Veterinary Medicine 93:82-91.

Ferris, D. A., R. A. Flores, C. W. Shanklin and M. K. Whitworth. 1995. Proximate analysis of food service wastes. *Appl. Engr. Agric.* 11:567-572.

Finstein, M.S. 1992. Composting in the Context of Municipal Solid Waste Management. Environmental Microbiology. Wiley-Liss, Inc. pp. 355-374.

Food Additive Petitions. Title 21 Code of Federal Regulations ' 571. U.S. Government Printing Office.

Food Additives. Title 21 Code of Federal Regulations ' 570. U.S. Government Printing Office.

Franklin and Associates. 1998. Characterization of Municipal Solid Waste in the United States: 1997 Update. U.S.-Environmental Protection Agency. Municipal and Industrial Solid Waste Division. Office of Solid Waste. Report No. EPA530-R-98-007. Franklin Associates, Ltd. Prairie Village, KS.

Garbage Feeding Not All Waste. Comments by Dr. Geneva Acor, University of Florida, 1994, pg. 1.

Glenn, J. 1992a. I. The State of Garbage in America: Biocycle: *J. Waste. Recyc.* 33(4):46.

Glenn, J. 1992b. II. The State of Garbage in America: Biocycle: *J. Waste. Recyc.* 33(5):30.

Goering, H. K., and P. J. Van Soest. 1975. Forage fiber analyses (apparatus, reagents, procedures, and some applications). Agric. Handbook No. 379. ARS, USDA, Washington, DC.

Goldstein, J. 1995. Recycling food scraps into high end markets. Biocycle *J. Waste Recyc.* 36(8):40-42.

Grasser, L.A., J.G. Fadel, I. Garnett, and E.J. DePeters. 1995. Quantity and economic importance of nine selected by-products used in California dairy rations. *J. Dairy Sci.* 78:962-971.

Haque, A. K. M. A., J. J. Lyons and J. M. Vandepopuliere. 1991. Extrusion processing of broiler starter diets containing ground whole hens, poultry by-product meal, feather meal, or ground feathers. *Poul. Sci.* 70:234-240.

Harpster, H.W. 1997. Unusual silages based on food processing byproducts. *In*: Proc. Silage to Feedbunk-North American Conference, Feb. 11-13, Hershey, PA. NRAES Publ. 99.

Heitman, H., Jr., C. A. Perry, and L. K. Gamboa. 1956. Swine feeding experiments with cooked residential garbage. *J. Anim. Sci.* 15:1072-1077.

Henry, W. A. and F. B. Morrison. 1920. Feeds and Feeding. Seventeenth Revised Ed. The Henry-Morrison Co. Madison, WI.

Hill, B. and Wright, HF. 1992. Acute interstitial pneumonia in cattle associated with consumption of mould-damaged sweetpotatoes (Ipomoea batatas). *Aust Vet J* 60(2): 36-37.

Hinders, R. 1995. Cornell system useful in evaluation of rations containing byproducts. *Feedstuffs* 67(47):12.

Hodge J. E. 1953. Enzymatic Browning. *J. Agr. Food Chem.* 1:928-943.

House Reports vol. 19, 13377, 1980. House of Representatives.

House, J. A. and C. A. House. 1992. Vesicular Diseases, in Diseases if Swine, 7th edition; Iowa State University Press.

Hunter, J. M. 1919. II. Garbage as a hog feed. PP. 109-112. Fortieth Annual Report of the New Jersey Agricultural Experiment Station. Trenton, NJ.

Hurrell, R. F. 1990. Influence of the mallard reaction on the nutritional value of foods. Pages 245-258. *In*: The Maillard Reaction in Food Processing, Human Nutrition and Physiology. Birkhanser Verlag, Basel, Switzerland.

Jacob, M. T. 1993. Classifying the Supermarket Food Waste Stream. Biocycle: *J. Waste Recyc.* 34(2):46.

Kantor, L. S., K. Lipton, A. Manchester, and V. Oliveira. 1997. Estimating and Addressing America's Food Losses. Economic Research Service, U.S. Department of Agriculture Report. Food Review, Vol. 20, No.1. Jan-Apr. 1997, pages 3-11.

Koch, A. R. 1964. The New Jersey swine industry - an economic analysis. New Jersey Agricultural Experiment Station Mimeo A. E. 297. Rutgers, The State University of New Jersey. New Brunswick, NJ.

Kornegay, E. T., G. W. Vander Noot, W. S. MacGrath, and K. M. Barth. 1968. Nutritive Value of Garbage as a Feed for Swine. III. Vitamin Composition, Digestibility and Nitrogen Utilization of Various Types. *J. Anim. Sci.* 27:1345-1349.

Kornegay, E. T. and G. W. Vander Noot. 1968. Performance, digestibility of diet constituents and N-retention of swine fed diets with added water. *J. Anim. Sci.* 27:1307-1311.

Kornegay, E. T., G. W. Vander Noot, K. M. Barth, W. S. MacGrath, J. G. Welch, and E. D. Purkhiser. 1965. Nutritive value of garbage as a feed for swine. I. Chemical composition, digestibility and nitrogen utilization of various types of garbage. *J. Anim. Sci.* 24:319-324.

Kornegay, E. T., G. W. Vander Noot, K. M. Barth, G. Graber, W. S. MacGrath, R. L. Gilbreath, and F. J. Bielk. 1970. Nutritive evaluation of garbage as a feed for swine. Bull. No. 829. College of Agric. Environmental Sci. New Jersey Agric. Exp. Sta. Rutgers, State Univ. of New Jersey, New Brunswick.

Kummer, F. 1943. Equipment for Shredding Sweet Potatoes Prior to Drying for Livestock Feed. Auburn, AL, Agricultural Experiment Station of the Alabama Polytechnic Institute Circular no 89 page 1-13.

Law, J. 1915. History of foot-and-mouth disease. *Cornell Vet* 4:224.

Leman, A.D., B.E. Strew, W.L. Mengeling, S.D'Allaire, and D.J. Talyor 1992. Diseases of Swine; Iowa State University Press, 7th edition.

Lencki, R.W. 1995. Issues and Solutions for Recycling Food Wastes. Symposium - Recycled Feeds for Livestock and Poultry. March, 1995. Ontario, Canada.

Lovatt, J., A. N. Worden, J. Pickup, and C. E. Brett. 1943. The fattening of pigs on swill alone: a municipal enterprise. *Empire J. Exp. Agr.* 11:182.

Lyons, J. J. and J. M. Vandepopuliere. 1996. Spent leghorn hens converted into a feedstuff. *J. Appl. Poultry Res.* 5:18-25.

Mata-Alvarez, J., F. Cecchi, P. Llabres, and P. Pavan. 1992. Anaerobic Digestion of the Barcelona Central Food Market Organic Wastes. Plant Design and Feasibility Study. *Biores. Tech.* 42:33.

Mauer, F. 1975. African Swine Fever, in Diseases of Swine, 4th edition; Iowa State University Press.

Maynard et al. 1979. Animal Nutrition. Enzymatic Browning and the Maillard Reaction. 9th ed. McGraw-Hill Book Company. New York, NY.

McCarthy, J.E. 1992. Solid Waste: RCRA Reauthorization Issues. IB92018. Congressional Research Service. The Library of Congress. Washington, DC.

McClure, K. E., E. W. Klosterman, and R. R. Johnson. 1970. Feeding garbage to cattle and sheep. Ohio Report on Research and Development in Agriculture, Home Economics and Natural Resources. Volume 55:78-79. Ohio Agric. Res. Dev. Center, Wooster.

Mehren, M. 1996. Feeding Cull Vegetables and Fruit to Growing/Finish Cattle. Feeding Cull Vegetables and Fruit to Growing Cattle, Scottsdale, AZ, Academy of Veterinary Consultants Proceedings Page 1-6.

Meilgaard, M., G. V. Civille, and B. T. Carr. 1991. Sensory Evaluation Techniques (2nd. Rev. Ed.). CRC Press, Boca Raton, FL.

Mexico-United States Commission for the Prevention of Foot-and-Mouth Disease 1972. (NO FURTHER CITATION AVAILABLE FROM BOOK)

Minkler, F. C. 1914. Hog cholera and swine production. Circular No. 40. New Jersey Agricultural Experiment Station. Trenton, NJ.

Modebe, A. N. A. 1963. The value of African-type swill for pig feeding. *J. W. African Sci. Assn.* 8:33.

Mohler, J.R. and Traum, J., 1942. Foot-and-mouth disease. Separate No. 1882, Keeping Livestock Healthy. Yearbook of Agriculture, USDA, p. 263.

Mohler, J.R. 1929. The 1929 outbreak of foot-and-mouth disease in California. *J. Am. Vet. Med. Assoc.* 75:309.

Morrison, F. 1946. Roots, tubers and miscellaneous forages. Feeds and Feeding. Ithaca, NY, The Morrison Publishing Company, page 312-315.

Mulhern, F.J., 1953. Present status of vesicular exanthema eradication program. Proc. 57th Ann Meet. US. Livesto. Sanit Assoc. pp. 326-333.

Myer, R. O., T. A. DeBusk, J. H. Brendemuhl, and M. E. Rivas. 1994. Initial Assessment of Dehydrated Edible Restaurant Waste (DERW) as a Potential Feedstuff for Swine. Res. Rep. Al-1994-2. College of Agriculture. Florida Agricultural Experiment Station. University of Florida. Gainesville, Fl. pp. 44-51.

Myer, R. O., D. D. Johnson, W. S. Otwell and W. R. Walker. 1987. Evaluation of extruded scallop viscera-soy mixtures in diets for growing-finishing swine. *J. Anim. Sci.* 65 (suppl. 1): 33 (Abstract).

Myer, R. O. 1998. Evaluation of a dehydrated poultry (broiler) mortality - soybean meal product as a potential supplemental protein source for pig diets. Proc. Internat'l Conference on Animal Production Systems and the Environment. College of Agric., Iowa State Univ., Ames. pp. 85-90.

Myer, R. O., J. H. Brendemuhl, and D. D. Johnson. 1999. Evaluation of dehydrated restaurant food waste products as feedstuffs for finishing pigs. *J. Anim. Sci.* 77:685.

NCDA. 1996. Marketing North Carolina Sweetpotatoes Including Louisiana., North Carolina Department of Agriculture and Consumer Services, Raleigh, NC.

NCDA. 1998. North Carolina 1998 Agricultural Statistics. North Carolina Department of Agriculture and Consumer Services, Raleigh, NC. Publication No. 190.

NJAES. 1919. II. Garbage as a Hog Feed. Fortieth Annual Report of the New Jersey Agricultural Experiment Station. Trenton, NJ.

NRC (National Research Council). 1982. United States-Canadian tables of feed composition. (3rd Ed.) National Academy Press, Washington, DC.

NRC. 1983. Underutilized Resources as Animal Feedstuffs. National Research Council. Washington, D.C. National Academy Press.

NRC. 1980. Mineral Tolerance of Domestic Animals. National Academy Press, Washington, D.C.

NRC. 1998. Nutrient Requirements of Swine (10th Rev. Ed). National Academy Press, Washington, D.C.

Passell, P. 1991. The Garbage Problem: It may be Politics, Not Nature. New York Times. February 26, 1991. pp. C1, C6.

Peckham, J., F. E. Mitchell, O. H. Jones, Jr., and B. Doupnik, Jr. 1972. Atypical interstitial pneumonia in cattle fed moldy sweet potatoes. *JAVMA* 160(2): 169-172.

Peirce, L. 1987. Tuber and Tuberous Rooted Crops. Vegetables: Characteristics, Production, and Marketing. New York, John Wiley and Sons, page 287-308.

Peterson, L.A. 1957. Growth and carcass comparisons of swine fed a concentrate ration, cooked garbage, and additional protein, vitamin and mineral supplements. M.S. Thesis. Univ. of Conn., Storrs.

Pettigrew, J. E., Jr. and R. E. Moser. 1991. Fat in swine nutrition. Page 133-146 in Swine Nutrition. E. R. Miller, D. E. Ullrey, and A. J. Lewis, eds. Butterworth-Heinemann. Stoneham, U.K.

Pigott, G. M. 1981. Seafood waste management in the Northwest and Alaska. Report No. 40. In W. S. Otwell (Ed.). Seafood Waste Management in the 1980s:

Conference Proceedings. Sea Grant College Program, University of Florida, Gainesville.

Polanski, J. 1996. Legalizing the feeding of nonmeat food wastes to livestock. *In*: Proc. of Food Waste Recycling Symposium. M. Westendorf and E. Zirkle, (Eds). USDA, Rutgers Univ., and New Jersey Dept. of Ag. Jan. 22,23. Atlantic City, NJ. pp.3-9.

Polanski, J. 1995. Legalizing the Feeding of Nonmeat Food Wastes to Livestock. Appl. Engr. Agr. 11(1):115.

Pond, W. G. And J. H. Maner. 1984. Swine Production and Nutrition. AVI Publishing Co., Inc., Westport, CT. pp. 336-368.

Pond, W. G., D. C. Church, and K. R. Pond. 1995. Basic Animal Nutrition and Feeding (4th Rev. Ed.). John Wiley & Sons. New York.

Poore, M. H., G. M. Gregory, J.L. Hart, and P.R. Ferket. 1998. Value of alternative ingredients in calf growing rations. *J. Anim. Sci.* 76 (Suppl. 1): 304 (abstr).

Poultry Improvement Plan. 1998. Title 9 Code of Federal Regulations. 145 and 147. U.S. Government Printing Office.

Poultry Products Inspection Act (21 U.S.C. 451 et seq.).

Price, A. T., D. A. Derr, J. L. Suhr and A. J. Higgins. 1985. Food Waste Recycling Through Swine. Biocycle: *J. Waste. Recyc.* 26(2):34-37.

Prokop, W. H. 1996. The rendering industry - a commitment to public service. D. A. Franco and W. Swanson, Eds. The Original Recyclers. National Renderers Association. Merrifield, VA.

Public Health Service Act. (42 U.S.C. ' 201 et seq.).

Public Law 96-468-October 17, 1980. Also referred to as the "Swine Health Protection Act" and also contained in the Federal Code of Regulations, Sub-Chapter L-Swine Health Protection, Part 166.

Radostits, O., D. C. Blood, and C. C. Gay. 1994. Veterinary Medicine A Textbook of the Diseases of Cattle, Sheep, Pigs, Goats and Horses. Bailliere Tindall, London, UK, page 1617-1621.

Ribeiro, J. M. and R. J. Azevedo. 1961. Reapparition de la peste porcine (P.P.A.) au Portugal. *Bull. Off. Intern. Epizoot.* 55:88.

Ribeiro, J. M., R. J. Azevedo, M. J. O. Teixeiro, M. C. Braco Forte, R. Rodriguez, A. M. Ribeiro, E. Oliveiro, F. Noronha, C. Grave Pereira, and J. Dias Vigario, 1958. Peste porcine provoquee par une souche differente (souche L)de la souche classique. *Bull. Off. Intern. Epizoot.* 50:516.

Rivas, M. E., J. H. Brendemuhl, D. D. Johnson, and R. O. Myer. 1994. Digestibility by Swine and Microbiological Assessment of Dehydrated Edible Restaurant Waste. Res. Rep. Al-1994-3. College of Agriculture. Florida Agricultural Experiment Station. University of Florida. Gainesville, FL.

Rivas, M. E., J. H. Brendemuhl, R. O. Myer, and D. D. Johnson. 1995. Chemical composition and digestibility of dehydrated edible restaurant waste (DERW) as a feedstuff for swine. *J. Anim. Sci.* 73 (suppl. 1): 177 (Abstract).

Rogers, G. M., M. H. Poore, B. L. Ferko, T. T. Brown, T. G. Deaton, and J. W. Bawden. 1999. Dental wear and growth performance in steers fed sweetpotato cannery waste. *JAVMA* 214:681.

Rogers, P. A. M. and D. B. R. Poole. 1987. Incisor wear in cattle on self-fed silage. *Vet. Rec.* 120:348.

Rogers, G., and M. H. Poore. 1997. Dental effects of feeding sweet potato cannery waste in beef cattle. *Compendium on Cont Ed for Practicing Vet* 19:541-546.

Rogers, G.M. and M.H. Poore. 1994. Alternative feeds for reducing beef cow feed costs. *Vet. Med.* 89(11):1073-1084.

Rogers, G., M.H. Poore, B.L. Ferko, R.P. Kusy, T.G. Deaton, and J.W. Bawden. 1997. In vitro effects of an acidic by-product feed on bovine teeth. *AJVR* 58:498-503.

Rolston, L. H., C.A. Clark, J.M. Cannon, W.M. Randle, E.G. Riley, P.W. Wilson, and M.L. Robbins. Beauregard sweet potato. *HortScience* 22:1338-1339.

Ruiz, M. 1982. Sweet Potatoes (Ipomoea batatas (L) Lam) for Beef Production: Agronomic and Conservation Aspects and Animal Responses. Sweet Potato Proceedings of the First International Symposium, Shanhua, Tainan, Taiwan, Asian Vegetable Research and Development Center.

Sauter, E., D. D. Hinman, and J. F. Parkinson. 1985. The lactic acid and volatile fatty acid content and in vitro organic matter digestibility of silages made from potato processing residues and barley straw. *J. Anim. Sci.* 60: 1087-1094.

Schad, G.A., C.H. Duffy, D.A. Leiby, K.D. Murrell, and E.W. Zirkle. 1987. *Trichinella Spiralis* in an Agricultural Ecosystem: Transmission under Natural and Experimentally Modified On-Farm Conditions. *J. Parasit.* 73(1):95.

Seath, D., L. L. Rusoff, G. D. Miller, and C. Branton. 1947. Utilizing Sweet Potatoes as Feed for Dairy Cattle., Louisiana Agricultural Experiment Station, Bulletin #423.

Singletary, C., S. E. McCraine, and L. Berwick. 1950. Dehydrated Sweet Potato Meal for Fattening Steers., Louisiana State University and Agricultural and Mechanical College Bulletin #446, page 3-7.

Sistrunk, W. and I. K. Karim. 1977. Disposal of lye-peeling wastes from sweet potatoes by fermentation for livestock feed. *Arkansas Farm Res.* 26(1): 8-9.

Shanahan, M. S. 1954. Present situation on foot-and-mouth disease. *Mil. Surg* 114:444. USDA Release, 1947. Summary developments in the Mexican outbreak of foot-and-mouth disease. Jan 28, 1947.

Skajaa, J. 1989. Food Waste Recycling in Denmark. Biocycle: *J. Waste Recyc.* 30(11):70.

Stack, Charles R. and Prasad S. Kodukula (April 24,1995); Production of Food Processing Biosolids and Their Use as Animal Feed: An Overview; Presentation to the AAFCO Environmental Issues Committee; Indianapolis, Indiana.

State of New Jersey. 1983. Right-to-Farm Act. Chapter 31, P.L. 1983. Approved January 26, 1983. N.J.S.A. 4:1c-1 et al.

Steele, J.H. 1968. Garbage-Borne Diseases in Swine. Proc. 72nd Ann. Meet. U.S. Livestock Sanitation Assoc. Appendix I.

Steele, J.H. and L.A. Busch. 1968. Garbage-Borne Diseases of Swine and their Relationship to Public Health. Proc. 72nd Ann. Meet. U.S. Livestock Sanitation Assoc. Page 335-342.

Substances That are Generally Recognized as Safe. Title 21 Code of Federal Regulations 582.1. U.S. Government Printing Office.

Suhr, J. L., A. T. Price, D. A. Derr, and A. J. Higgins. 1984. Feasibility of Food Waste Recycling in New Jersey. Report P-03545-01-84 Cook College. New Jersey Agricultural Experiment Station. Rutgers—the State University of New Jersey. New Brunswick, New Jersey.

Sulfiting Agents; Revocation of GRAS Status for Use on Fruits and Vegetables to be Served or Sold Raw to Consumers. 51 Federal Register 25021, July 9, 1986.

Swine Health Protection Act. 1998. Title 9 Code of Federal Regulations ' 166. U.S. Government Printing Office.

Swine Health Protection, Definitions in Alphabetical Order. 1998. Title 9 Code of Federal Regulations '166.1. U.S. Government Printing Office.

Swope, R.L., H.W. Harpster, R.S. Kensinger, and V.H. Baumer. 1995. Nutritive value of human infant formulas for young calves. *J. Anim. Sci.* 73(Suppl 1): 249 (Abstr.).

Tacon, A. G. J. and A. J. Jackson. 1985. Utilization of conventional and unconventional protein sources in practical fish feeds. *In*: Nutrition and Feeding in Fish. (C. B. Cowey, A. M.Mackie and J. B. Bell, Eds.). Academic Press, New York, NY. pp. 118-145.

Tadtiyanant, C., J. J. Lyons and J. M. Vandepopuliere. 1993. Extrusion processing used to convert dead poultry, feathers, eggshells, hatchery waste, and mechanically deboned residue into feedstuffs for poultry. *Poul. Sci.* 72:1515-1527.

Tadtiyanant, C., J. J. Lyons, and J. M. Vandepopuliere. 1989. Utilization of extruded poultry farm mortalities and feathers in broiler starter diets. *Poultry Sci.* 68(Suppl. 1):145 (Abstr.).

The State of Garbage: The Northeast-Midwest Leads the Nation in Recycling. Northeast Midwest Economic Review, Vol. 8, No. 6, July 1995.

Tolan, A. 1983. Sources of food waste, UK and European aspects. Page 15-27. In: D. A. Ledward, A. J. Taylor, and R. A. Lawrie (Eds.). Upgrading Waste for Feed and Food. Butterworths, London.

U.S. Congress. 1980. Swine Health Protection Act. Public Law 96-468.

U.S. verses An Article of Food *** Coco Rico, 752 F.2d 11 (1ˢᵗ Cir. 1985).

United States v. 50 Boxes * * * Cafergot P- B Suppositories, 721 F.Supp. 1462, 1465 (D. Mass. 1989), aff'd, 909 F.2d 24 (1st Cir. 1990); An Article of Drug * * * Furestrol Vaginal Suppositories, 251 F.Supp. 1307 (N.D. Ga. 1968), aff'd, 415 F.2d 390 (5th Cir. 1969), see also 5,906 Boxes, 745 F. 2d at 119 n.22.

USDA Release, 1954a. No. 999-54, April 14, 1954.

USDA Release, 1954b. No. 329-54. Dec. 22, 1954.

USDA Release, 1953. No. 1279-53. May 28, 1953.

USDA. 1996. Nutrient Database for Standard Reference, Release 11, Sweetpotato, Raw.

USDA-APHIS, CEAH. 1995. "Risk Assessment of the Practice of Feeding Recycled Commodities to Domesticated Swine in the U.S.", page D-4 to D-6, January 1995.

USDA-APHIS, VS. 1995. Risk of feeding food waste to swine: Public health diseases. Centers for Epidemiology and Animal Health Bulletin. April 1995. Fort Collins, CO.

USDA-APHIS, VS 1990. Heat-Treating Food Waste—Equipment and Methods. USDA Animal and Plant Health Inspection Service, Veterinary Services. Program Aid No. 1324.

USDA-APHIS, VS. 1995. Swine waste feeder profile reveals types, sources, amounts, and risks of waste fed. United States Department of Agriculture Animal and Plant Health Inspection Service, Veterinary Services. Fort Collins, CO.

USDA-APHIS, VS. 1995. Risk Assessment of the Practice of Feeding Recycled Commodities to Domesticated Swine in the U.S. United States Department of Agriculture Animal and Plant Health Inspection Service, Veterinary Services. Centers for Epidemiology and Animal Health. Fort Collins, CO.

USDA-APHIS, VS 1990. Heat-Treating Food Waste—Equipment and Methods. USDA Animal and Plant Health Inspection Service, Veterinary Services. Program Aid No. 1324.

USDA-APHIS. 1984. Swine Health Protection. USDA Animal and Plant Health Inspection Service. *Federal Register.* 49(72):14495-14497.

USDA-APHIS. 1982a. Swine Health Protection Provisions. USDA Animal and Plant Health Inspection Service. *Federal Register.* 47(213):49940-49948.

USDA-APHIS. 1982b. State Status Regarding Enforcement of the Swine Health Protection Act. USDA Animal and Plant Health Inspection Service. *Federal Register.* 47(251):58217-58218.

USDA-ERS. 1992. Animal Feeds Compendium. M. S. Ash, Ed. USDA-Economics Research Service. Agricultural Economic Report Number 656. PP. 65-110.

Van Oirschot, J. T. 1992. Hog Cholera, in Diseases of Swine, 7th edition.

Van Loon, Dirk, Small-Scale Pig Raising, A Garden Way Publishing Book, pg. 123.

Wagner, Matina (October 1995); Resource Recycling, Solid Waste Association of North America (SWANA), Silver Spring, Maryland.

Walker, P. M. and A. E. Wertz. 1994. Analysis of selected fractionates of a pulped food waste and dish water slurry combination collected from university cafeterias. *J. Anim. Sci.* 72 (suppl. 1): 137 (abstract).

Walker, P. and T. Kelly. 1998. Selected fractionate composition and microbiological analysis of institutional food waste pre- and post-extrusion. In Proc. 2nd. Food Waste Recycling Symposium. (Westendorf and Zirkle, Ed.) New Jersey Department of Agriculture and Rutgers Cooperative Extension. Trenton and New Brunswick, NJ.

Walker, P. M. And T. Kelly. 1997. Selected fractionate composition and microbiological analysis of institutional food waste, pre- and post-extrusion. Proc. Food Waste Recycling Sym. Rutgers Coop. Ext., Rutgers Univ. - Cook College, New Brunswick, NJ. pp. 61-72.

Walter, Bill. Nutrient Content of Sweetpotatoes (July 10,1997), Telephone conversation.

Waterman, J. J. 1975. Measures, stowage rates and yields of fishery products. Torry advisory note no. 17. Torry Research Station. Aberdeen, Scotland.

Weinberger verses Hynson, Wescott and Dunning Inc., 412 U.S. 606, 632 (1073).

West, R. L. and R. O. Myer. 1987. Carcass and meat quality characteristics of swine as affected by the consumption of peanuts remaining in the field after harvest. *J. Anim. Sci.* 65:475-480.

Westendorf, M. L., T. Schuler, and E. W. Zirkle. 1999. Nutritional quality of recycled food plate waste in diets fed to swine. *Prof. Anim. Sci.* 15(2):106-111.

Westendorf, M. L. 1996. The use of food waste as a feedstuff in swine diets. Proc. Food Waste Recycling Sym. Rutgers Coop. Ext., Rutgers Univ. - Cook College, New Brunswick, NJ. pp. 24-32.

Westendorf, M. L., Z. C. Dong, and P. A. Schoknecht. 1998. Recycled cafeteria food waste as a feed for swine: nutrient content, digestibility, growth, and meat quality. *J. Anim. Sci.* 76:3250.

Westendorf, M. and E. Zirkle (Ed.). 1996. Proc. of Food Waste Recycling Symposium. USDA, Rutgers Univ. and New Jersey Dept. of Ag. Jan. 22-23, Atlantic City, NJ.

Westendorf, M. L., E. W. Zirkle, and R. Gordon. 1996. Feeding food or table waste to livestock. *Prof. Anim. Sci.* 12(3):129-137.

Wilson, L. L. and P. G. Lemieux. 1980. Factory canning and food processing wastes as feedstuffs and fertilizers. *In:* M. Bewick (Ed.) Handbook of Organic Waste Conversion. pp. 243-267. VanNostrand Reinhold Co., New York.

Wilson, B., D. T. C. Yang and M. R. Boyd. 1970. Toxicity of mould-damaged sweet potatoes (Ipomoea batatas). Nature 227: 521-522.

Wilson, L. and W. W. Collins. 1993. What is the difference between a sweetpotato and a yam? North Carolina Cooperative Extension Service.

Wilson, J. H., R. J. MacKay, H. Nguyen, W. S. Cripe. 1981. Atypical Interstitial Pneumonia: Moldy Sweet Potato Poisoning in a Florida Beef Herd. *Florida Veterinary Journal* 10(Fall 1981): 16-17.

Wilson, B. 1973. Toxicity of mold-damaged sweetpotatoes. Nutrition Reviews 31(3): 73-78.

Wohlt, J. E., J. Petro, G. M. J. Horton, R. L. Gilbreath, and S. M. Tweed. 1994. Composition, Preservation, and Use of Sea Clam Viscera as a Protein Supplement for Growing Pigs. *J. Anim. Sci.* 72:546.

Woolfe, J. 1992. Livestock Feeding with Roots and Tubers. Sweetpotato an Untapped Food Resource. Cambridge University Press.

Wu, J. 1980. Energy value of sweet potato chips for young swine. *J. Anim. Sci.* 51(6): 1261-1265.

Zhang, Y. and C. M. Parsons. 1996. Effects of overprocessing on the nutritional quality of peanut meal. *Poul. Sci.* 75:514-518.

Index

ISBN 0-8138-2540-7

90000